B-36 PEACEMAKER

The Big Stick of Strategic Air Command

H.J. CAMPBELL

Electrikbooks

First Edition Copyright © 2021, 2022 by Harold J Campbell

Second Edition Copyright © 2024 by Harold J Campbell

All rights reserved. No part of this publication may be reproduced or utilized in any form or by any means electronic, mechanical, photocopy, recording, or otherwise, stored in an information and retrieval system, or transmitted in any form without express written permission of the publisher.

This 2nd Edition was updated to replace the cover, add photos, enlarge existing photos, reformat lineage chart, reformat tail number list, and add Colors and Markings appendix.

This book was updated on 13 July 2025 to replace the cover and make minor formatting changes only.

On the Cover: A B-36 takes off. (U.S. Air Force)

Table of Contents

Introduction iii
Design and Development 1
Roles and Missions 13
    Strategic Bombardment 13
    Revolt of the Admirals 14
    Strategic Reconnaissance 21
Models and Variants 24
    Prototypes 31
    Bombers 36
    Strategic Reconnaissance 44
    Special Mission 48
    Experiments 50
    Featherweights 56
Organization and Basing 58
Firsts and Records 69
Operations 72
    Operational Missions 72
    Operational Deployments 80
    Competitions and Exercises 85
    Atomic Tests 88
    Texas Twister 94
    Broken Arrow 97
Displays 99
    B-36J (52-2827) 99
    B-36J (52-2220) 100
    B-36J (52-2217) 101

RB-36H (51-13730)	101
Appendix A: Specifications	103
Appendix B: Colors and Markings	115
Appendix C: Tail Numbers	121
Bibliography	i
Notes and Citations	ii
About the Author	v

Introduction

Theodore Roosevelt once said, referring to foreign policy, "Speak softly and carry a big stick", which was the essence of Strategic Air Command's big stick, the B-36 Peacemaker. The B-36 was the mainstay of United States strategic deterrence policy during the early years of the Cold War. This policy was based on the theory of Massive Retaliation, in which an attack by the Soviet Union on the United States or its allies would be met with a full retaliatory strike on the Soviet Union. The purpose was to establish a nuclear deterrence through strength that would prevent war from breaking out. The B-36 was the big stick used to enforce this policy.

The B-36 seemed ready-made for this role. But it was simply in the right place at the right time. The United States Army Air Corps began its quest for a "Giant Bomber" in the mid-1930s under a project called Bomber, Long Range. Their goal was to develop an ultra-long-range bomber capable of carrying a large bombload and to determine the maximum range these aircraft could achieve. This project resulted in two of the largest American aircraft built at that time.

By 1941, Nazi forces had swept across Europe and occupied its major capitals including Paris. They were now poised to overtake Great Britain. Meanwhile, Japan had overtaken much of China and the Western Pacific. Air Corps leaders were concerned that the United States would be forced into the war and would need to engage the enemy from bases in the United States. This concern led them to step up their quest for a giant bomber to a whole new level, and ultimately led to development of the B-36 Peacemaker. Although it never saw service during World War II (WWII) due to higher priorities, B-36 development continued, and the XB-36 made its first flight in August 1946 just days before Japan's surrender from WWII.

This book tells the story of the Peacemaker's transformation from an aircraft without a mission and its rise to Strategic Air Command's big stick in the early Cold War years. It faced great scrutiny from within the newly formed United States Air Force and the Department of Defense. It was nearly cancelled several times, but ultimately proved its worth as the only weapon system capable of carrying the atomic bomb to far-away targets. Though it was eventually replaced by faster aircraft capable of worldwide range through aerial refueling, its place in airpower history is well established.

Design and Development

By the mid-1950s, the United States had grown into the world's dominant nuclear superpower. Its nuclear weapon stockpile had rapidly grown to over 2,400 weapons. Although the Soviet Union stockpile had only reached about 200 weapons, United States planners feared that its aggressiveness, demonstrated by its domination of Eastern Europe, would result in an offensive strike against the United States or its allies in Western Europe. In 1954, President Eisenhower endorsed the strategy of "Massive Retaliation" in which any attack by the Soviet Union on the United States, its territories and possessions, or its allies would result in a massive nuclear retaliation against the Soviet Union. The B-36 was now the nation's primary nuclear delivery platform and the mainstay of this strategy. Given its ability to carry the largest thermonuclear weapon in the arsenal, the 42,000-pound Mark-17 (Mk-17), the B-36 seemed purpose-built for the role. But it was simply in the right place at the right time. Its design started even before the United States entered World War II (WWII), long before the Cold War was imagined, and well before the advent of an air-droppable nuclear bomb.

The United States Army Air Corps began its quest for an ultra-long-range "Giant Bomber" in 1934 under the top-secret program called Project-A.[1] This project required a bomber capable of achieving a 5,000-mile range with a 2,000-pound bombload. Despite this range requirement, the stated mission of the bomber was strictly costal defense of the United States and protection of the Western Hemisphere including Alaska, Hawaii, and Panama. Both Boeing and Martin aircraft companies submitted designs, which were designated XB-15 and XB-16 respectively by the Air Corp. Both designs offered the largest aircraft ever built in the United States, although the Boeing design was smaller than the Martin design. After a review, the Air Corps authorized contracts on 12 May 1934 for prototypes of both designs. On 16 May, they revised the mission to include the destruction of distant land and naval targets, as well as the reinforcement of Alaska, Hawaii, and Panama without intermediate servicing facilities. Ultimately, the Air Corps selected the Boeing design, and an XB-15 prototype was completed in 1937.[2]

Project D, established in 1935, was a follow-on to Project A. Its goal was to study the maximum range that an ultra-long-range bomber could achieve. There were two major manufacturers selected under this program, Douglas Aircraft and Sikorsky. Both companies submitted design proposals and built mock-ups for evaluation. In the spring of 1935, the Air Corps decided to merge Project A and D into a single project called Bomber, Long Range. Under this program, the Boeing XB-15 design was given the designation XBLR1 for eXperimental Bomber, Long-Range, while the Douglas and Sikorsky designs were designated XBLR2 and XBLR3. In mid-1936, the Air Corps selected the Douglas Aircraft design. It was truly innovative and included an all-metal aircraft, and low-mounted wings with thick roots allowing mechanics to access the engines

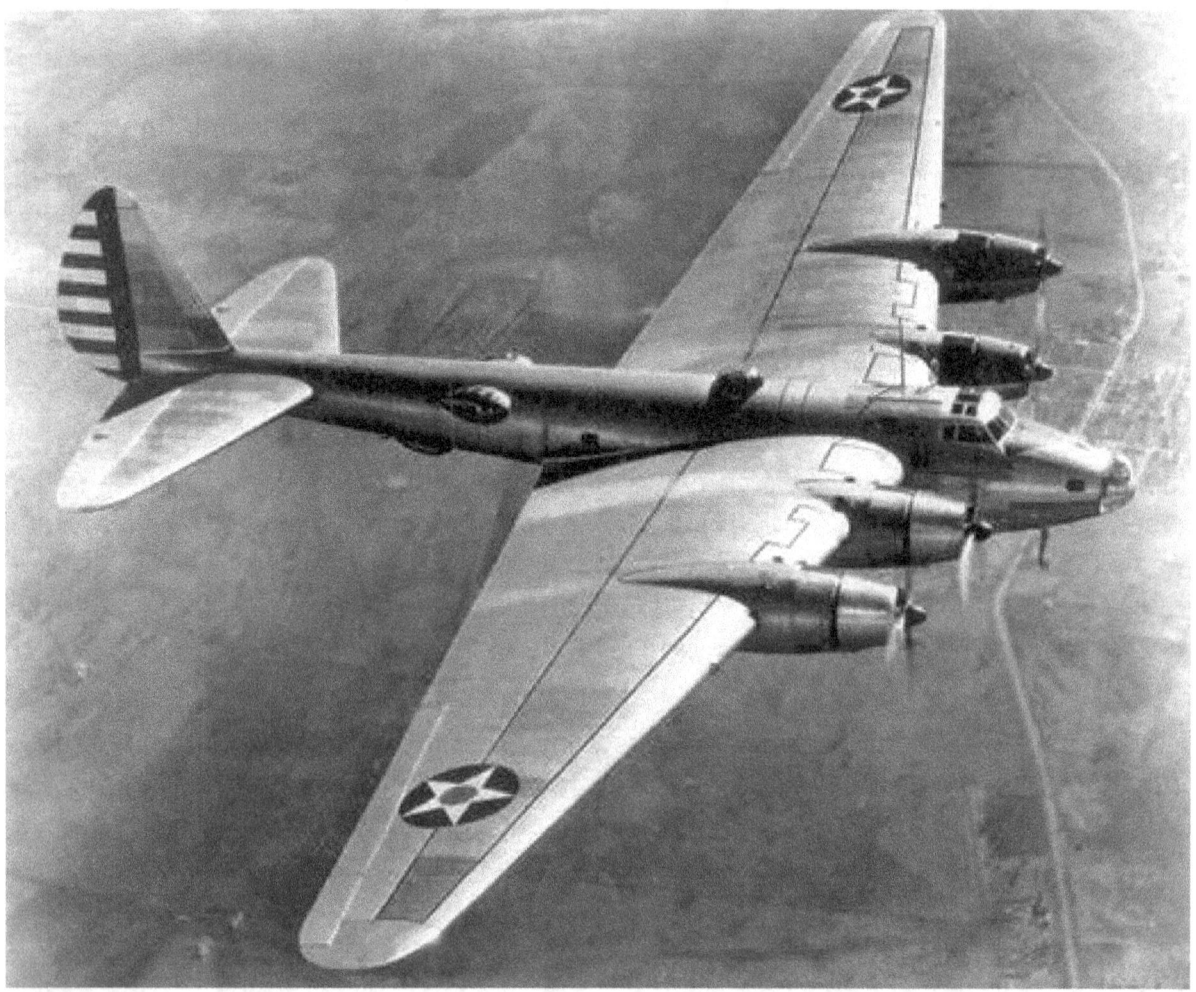

The Boeing XB-15 first flew on 15 October 1937 and was the first of the giant bomber experiments. (U.S. Air Force)

in flight. It featured four 1,600 horsepower XV-3420-1 Allison engines that were later replaced with 2,000 horsepower Wright R-3350-5 engines. By 1938, a production contract was established for the Douglas design, now redesignated B-19, and one XB-19 was ordered.

In 1939, war exploded in Europe when Nazi forces overran Poland in just twenty days during a "Blitzkrieg" campaign that marked the beginning of WWII. President Roosevelt was determined to keep the United States out of the war, and he adopted a strategy of supplying British forces with American made weapons. On 16 May 1940, in his address to Congress, he requested funding for the defense industry to gear up and produce 50,000 military aircraft to support the war effort and avoid America's inevitable involvement.

But by July 1940, Hitler had conquered most of Europe and the Nazis began attacking British shipping in the English Channel. By August the Luftwaffe was attacking the British Isles. The Royal Air Force managed to stave off the attack and prevent defeat, but American war planners feared that Britain might fall to Hitler. In April 1941, the London blitz was at its height. U-boats were slowly starving Britain's population and industry. The Nazis continued to conquer territory across Europe, and the United States War Department became increasingly concerned that Great Britain would be overtaken leaving the United States without land bases in Europe. Against this background President Roosevelt directed, in conference with Air Corps leaders, that an "intercontinental" heavy bomber, which "could carry the war to the prospective enemy from this side of the troubled water" should be developed.[3]

On 11 April 1941, the Air Corps issued a specification for the first intercontinental strategic bomber that took its quest for a Giant Bomber to a whole new level. The specification required a bomber capable of launching from the United States and reaching Europe non-stop. It specified a 12,000-mile range at 25,000 feet altitude, and 275 miles per hour (mph) cruise speed at 45,000 feet. Combat radius was specified at 5,000 miles carrying a 10,000-pound bombload. The expected maximum speed was 450-mph at 45,000 feet. The maximum bombload was 72,000 pounds.[4] This was before the advent of aerial refueling which made the feat even more challenging. By 3 May, Boeing, Consolidated Aircraft, and Douglas Aircraft submitted preliminary designs based on these specifications. Each of their designs fell short and the companies found it difficult to meet the specifications.

Meanwhile, on 27 June 1941, Douglas Aircraft made the first flight of the XB-19 which was the largest American aircraft ever built at the time. Its four 2,000 horsepower Wright R-3350-5 engines produced a maximum speed of 224 mph at 15,700 feet altitude. It had a 212-foot wingspan, 132-foot length, and 160,000 pounds gross weight. The aircraft carried a normal bombload of 18,700 pounds and had maximum capacity of 37,100 pounds with reduced fuel load. It had a range of 5,200 miles and the defensive armament included two 37mm cannons and five .50-caliber and six .30-caliber machine guns. It had a 16-man crew and featured sleeping and galley accommodations for a second crew to allow the aircraft to remain airborne for 24 hours. Although the aircraft was relatively successful and had few problems, only the one XB-19 was built. The aircraft was now considered mostly obsolete, and the Air Corps elected to use it as a flying laboratory to test large bomber concepts.

On 19 August 1941, after the preliminary designs submitted failed to fully meet the previous specifications, the Air Corps revised the specifications and now required a 10,000-mile range at 25,000 feet, and 240 to 300-mph cruise speed at 40,000 feet. Combat radius was specified at 4,000 miles carrying a 10,000-pound bombload. The expected

The Douglas XB-19 was the largest American aircraft ever built at that time when it made its first flight on 27 June 1941. (U.S. Air Force)

maximum speed was still 450-mph at 45,000 feet. The maximum bombload was 72,000 pounds.[5]

Given the size and range of modern-day aircraft, it is difficult to understand how far-sighted even these reduced specifications were at the time. It is important to understand the context of the day to fully comprehend the enormity and significance of the challenge. It was essentially like going to the moon was in the 1960s. As shown below, the

Characteristic	Specification	B-17G	XB-19A	B-29B	B-32
Design Initiated	December 1941	August 1934	October 1935	August 1940	June 1940
Range	10,000 miles	1,761 miles	4,200 miles	4,523 miles	3,800 miles
Cruise Speed	240 to 300 mph	197 mph	185 mph	220 mph	290 mph
Maximum Speed	450 mph	320 mph	265 mph	415 mph	357 mph
Bombload	10,000 pounds	10,000 pounds	18,700 pounds	10,000 pounds	10,000 pounds
Maximum Bombload	72,000 pounds	12,800 pounds	37,100 pounds	20,000 pounds	20,000 pounds
Service Ceiling	40,000 feet	36,450 feet	39,000 feet	41,400 feet	30,700 feet

The Air Corps specifications issued in 1941 represented a technological leap far beyond anything in existence and resulted in the largest piston engine aircraft ever produced.

specifications called for an aircraft far greater than anything in existence or even on the drawing boards.

The Air Corps sent invitations to Boeing and Consolidated Aircraft to bid on the revised specifications and present new preliminary designs for an intercontinental bomber. Boeing submitted their Model 384 and 385, but both were considered half-hearted attempts. Neither one met the requirements although the Model 385 came close to it. The Model 384 was a four-engine aircraft while the Model 385 used six engines.[6] The Air Corps believed that Boeing did not seriously consider the long-range bomber requirement and that their design was overly conservative. On the other hand, Consolidated Aircraft was keenly aware of the Air Corp's interest in a large bomber with extended range and had been working on several designs.[7] The company submitted their Model 35, which also included a four-engine and six-engine version, but the engines were mounted in a "pusher" (aft-facing) configuration instead the standard tractor configuration.

On 3 October 1941 after reviewing the submitted designs[8] the Army Air Forces (AAF), which had been established on 20 June 1941 with Major General H.H. "Hap" Arnold as its Chief, recommended further study of the Consolidated Aircraft designs. On 6 October, Consolidated Aircraft submitted a full proposal of $15 million for two prototype aircraft. They also proposed $800,000 in fixed fee and stipulated that the Government must provide no red tape or constantly changing directives. In return for these considerations, Consolidated Aircraft promised to deliver the first aircraft 30 months after receipt of order (ARO) in May 1944, and the second 36 months ARO in November 1944.

Brigadier General George Kenney, who then headed the Air Corps Experimental Division and Engineering School at Wright Field and would later become the first Commanding General of Strategic Air Command (SAC), took Consolidated Aircraft's drawings to Arnold. Kenney recommended, "That we go ahead on that design. Our evaluation at the Materiel Division indicated it was the best of the lot." Interestingly, Wright Field personnel, working independently of Consolidated Aircraft's engineers, also had come up with a six-engine pusher-type bomber for the intercontinental mission. It was not vastly different from the firm's proposed design.[9]

On 16 October 1941, Arnold approved the proposal and the AAF ordered the two prototypes on 15 November based on the 4-engine design. On 22 November 1941, the Engineering Division at Wright Field decided on the 6-engine rather than the 4-engine design and the AAF revised the contract. On 10 December 1941, Consolidated Aircraft redesignated the Model 35 as Model 36 so it aligned with the XB-36 designation and would not be confused with the Northrop XB-35 Flying Wing.

On 7 December 1941, Japan attacked Pearl Harbor and brought the United States into WWII. On 11 December 1941, Germany and Italy declared war on the United States. Suddenly, the United States was thrust into war on all continents. Facing the

B-36 PEACEMAKER: THE BIG STICK OF STRATEGIC AIR COMMAND

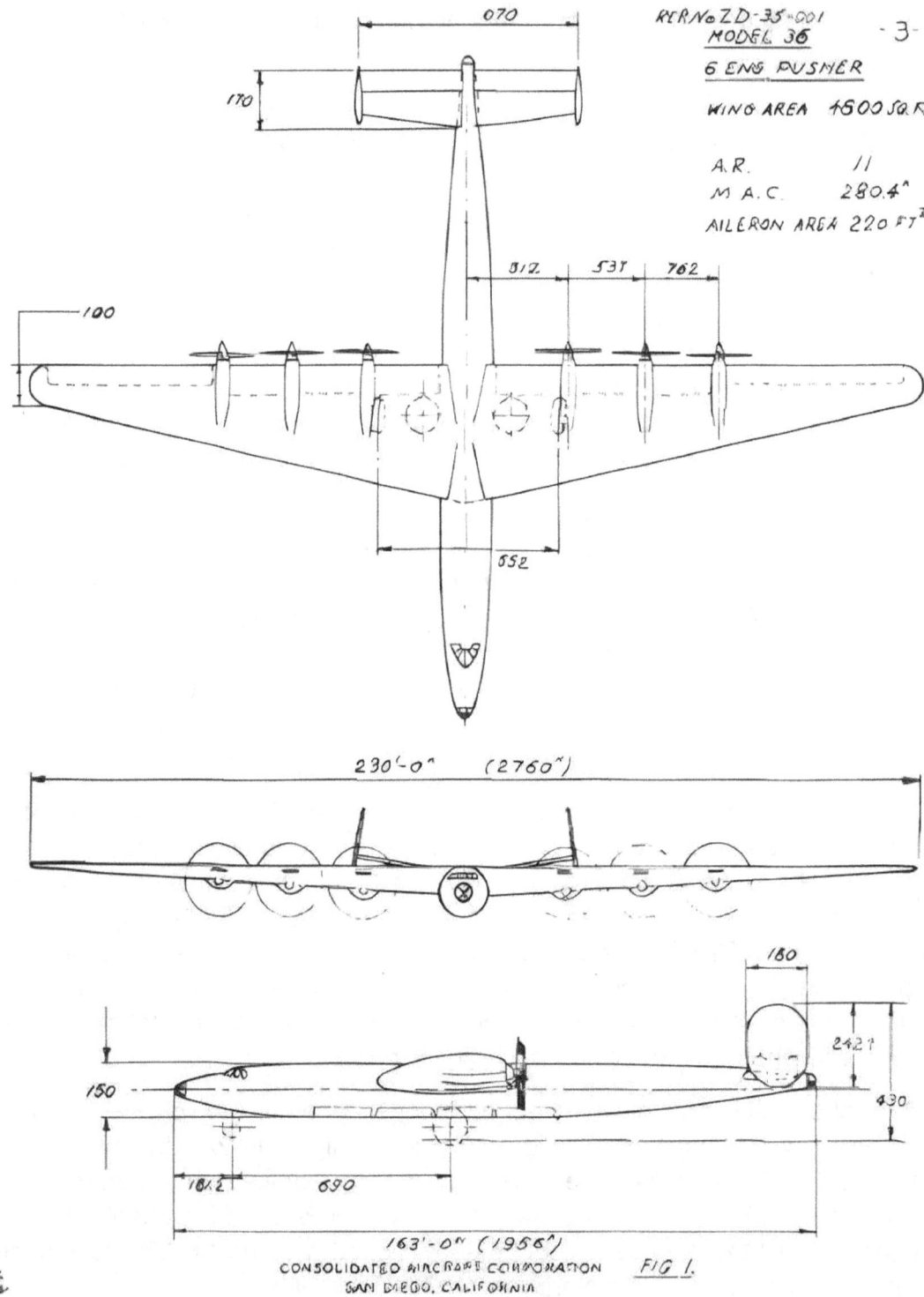

The Consolidated Model 36 evolved from the original Model 35 proposal into a six-engine pusher design with a twin tail that eventually was refined in the final XB-36 design. (Consolidated Aircraft Corporation)

possibility of becoming isolated, the importance of an intercontinental bomber became clear. The AAF placed high priority on the development of the aircraft that would become the B-36 Peacemaker. Not only could it reach Europe from United States bases, but it could reach Tokyo from American bases in Hawaii. This was especially important as Japan captured most Pacific islands and bases in China.

But, as the war raged in Europe and bomber bases in Great Britain remained protected, pressure was on for high-priority production of proven bomber aircraft such as the B-17 and B-24. The B-29 Superfortress also received top priority to support the deteriorating military situation in the Pacific theater. Consolidated Aircraft's facilities in San Diego were turned mainly to producing vitally needed B-24 Liberators, C-87s, and Navy patrol bombers.

Although XB-36 development continued, it received a low priority. The two B-36 prototypes were designated XB-36. The first XB-36 was an experimental proof-of-concept aircraft while the second aircraft was subsequently redesignated YB-36 and would be a more production representative aircraft. The experimental B-36 project presented many challenges for Consolidated Aircraft's design team. There was, at the time, little or no experience with the structure to gross weight ratios dictated by the plane's size. The engines to power the plane were still on paper at Pratt & Whitney (P&W). The 3,000-horsepower promised for each of the R-4360 Wasp Major engines was unheard of. The job progressed slowly.

The AAF broke ground on a new plant in Fort Worth, TX designated Air Force Plant 4 on 18 April 1941 with local newspaper publisher Amon Carter, who had pushed the Government to locate the plant there, and Major General Hap Arnold in attendance. The $50 million plant was to be operated by Consolidated Aircraft and used to support increased production of B-24 Liberators and B-32 Dominators for the war effort. Ultimately, the AAF and Consolidated Aircraft agreed to move B-36 production to the Fort Worth plant which would become the history-making "home" of the Peacemaker.

In the spring of 1943, with XB-36 developmental work grinding forward, the war in the Pacific was taking a downward turn and China was in danger of collapse. Once again, the AAF needed an intercontinental bomber to reach targets in Japan and the Far East. To meet this urgent need Convair, which had by then been formed by the merger of Consolidated Aircraft with Vultee Aircraft Corporation[10], proposed that as much as two years could be cut from development time if work were begun on production models even before the experimental planes had been rolled out. This was a revolutionary concept that had never been attempted.[11] Despite the financial risk, the AAF accepted the proposal and Secretary Stimson waived procurement procedures to allow production to begin before testing was completed. The AAF awarded Convair a letter of intent for 100 B-36 aircraft on 23 July 1943, and a formal letter contract on 23 August.

Air Force Plant 4

In 1940, President Roosevelt requested production of 50,000 military aircraft to support the British war effort. But the United States lacked the production facilities or capability to take on this huge endeavor. Roosevelt quickly addressed this shortfall by establishing the Defense Plant Corporation with authority and a dedicated budget to rapidly develop the facilities and capability needed.

The wealthy Fort Worth, TX newspaperman Amon G. Carter, and the Executive VP of the Chamber of Commerce William Holden, lobbied hard to bring one of these new production plants to their city. The city had identified a track of land on the shore of Lake Worth where Consolidated Aircraft was already landing Navy seaplanes constructed in its San Diego, CA plant as a mid-continent stop on their cross-country delivery flights. Carter, Holden, and Consolidated Aircraft's head Rueben Fleet appealed to the Government that this location would allow production of both Army and Navy aircraft at the same location. They also suggested building a bomber base across the field that would share runways with the bomber plant.

Air Force Plant 4 became the home of the Peacemaker and many other famous planes. (U.S. Air Force)

Fort Worth faced a significant challenge from the city of Tulsa, OK. The Government was keen on building a Consolidated Aircraft plant there to address economic issues and they reasoned that Texas already had sufficient defense investment. A North American Aviation plant had just been approved for nearby Grand Prairie. After much debate, a compromise was reached, and a Douglas plant was built in Tulsa while a Consolidated Aircraft plant was built in Fort Worth. The Government also accepted the suggestion to build a bomber base that would be called Fort Worth Army Airfield and later Carswell Air Force Base (AFB).

Air Force Plant 4 is still a Government Owned – Contractor Operated (GOCO) facility that was operated for most of its life by Convair, which later became a division of General Dynamics. Over the years, the plant produced thousands of premier military aircraft including the B-24, B-36, B-58, F-102, F-104, F-111, F-16, and F-22. It is currently operated by Lockheed Martin and is the United States production facility for the F-35 Joint Strike Fighter.

Over the next year United States and Allied troops battled their way across Europe as B-17 Flying Fortresses, B-24 Liberators, and Allied aircraft pounded targets in Germany and Eastern Europe. In the Pacific, United States Marines clawed their way onto beaches, fought through jungles, and battled across mountainous terrain as Naval aviation punished Japanese positions to capture dozens of remote islands and secure airfields for B-29 bombers to begin their devastating assault on the Japanese homeland.

By mid-1944, the Marianas islands of Guam, Saipan, and Tinian were now in United States hands, once again reducing the need for the B-36. Despite this, development work on the B-36 continued. The final contract for 100 aircraft was definitized on 19

August 1944. But the AAF removed the priority rating from the contract, effectively placing the B-29 and B-32 priority ahead of the B-36. The contract also incorporated Change 7, which replaced the original twin tail design with a single tail.

On 25 May 1945, the AAF reviewed aircraft production requirements and agreed to reduce it by 17 percent, which totaled 17,000 aircraft over 18 months. However, on 6 August, General of the Air Force Hap Arnold[12] approved the Air Staff recommendation to keep the B-36 program intact. Long range bombers were needed more than ever as proven by the cost in lives and material paid for capturing Pacific bases. The B-36 was also projected to be more economical than both the B-29 and B-32 and was only half the cost of the B-29 to operate. Lieutenant General Hoyt Vanderburg, then Assistant

The B-36 was the largest piston-engine American combat aircraft ever built and was only half the cost of the B-29 to operate. (U.S. Air Force)

Chief of Staff for Operations, Commitments, and Requirements, who would later become Air Force Chief of Staff, advocated for four very heavy bomb groups equipped with B-36 aircraft as part of the planned 70-group Air Force. Delivery of the first aircraft was set for August 1945. But the Pacific war ended six months earlier than expected – and six days before the rollout of the first B-36, its nose jacked up to lower its tail, which was too tall for the hangar door.[13]

As 1946 began, the United States was winding down from WWII. It had been only four months since the war ended with the surrender of Japan, but demobilization was in full progress. The armed forces had released more than five million men with another 300,000 troops arriving home from overseas each month. The AAF was down to 734,000 people and dropping steadily. Until peacetime requirements could be figured out, the AAF stopped procurement of aircraft. Modification work was also suspended, as General Arnold put it, to be sure the AAF was not "shoeing any more dead horses." United States industry was converting as fast as it could from wartime production to output of consumer goods, including automobiles. Defense plants were closing, and armament contracts were canceled.[14]

Almost as soon as the war was over, the Soviet Union consolidated its control in Eastern Europe and pressed for new advantages in Central Europe. Soviet leader Joseph Stalin redefined his adversary to be the United States and Western Europe. In a speech at Fulton, MO in March 1946, Churchill declared, "From Stettin in the Baltic to Trieste in the Adriatic, an iron curtain has descended across the continent." It was political dynamite. Truman, who was present for the speech, refused to comment on Churchill's proposal for a British American alliance. He denied that he had seen a copy of the speech in advance.[15]

The atomic bomb, as demonstrated over Hiroshima and Nagasaki in August of 1945, approached the "ultimate" in destructive power. It also raised strategic bombing to more overwhelming military importance than ever — a position, incidentally, that had been predicted by airpower leaders back through the decades. Consequently, if the B-36 had not already existed in test stage in the aborning postwar world, it would have had to be created.[16] Despite his previous commitment to stop aircraft procurement until requirements could be figured out, Arnold agreed to continue work on the B-36.

But, despite this widespread recognition of the "big picture," the B-36 experienced some rough going when it first began rolling off the assembly line. First, it had almost inevitable technical bugs. General George C. Kenney, close to the B-36 from its inception and initially an advocate, expressed serious doubts about the plane shortly after it entered the aerial testing phase. On 12 December 1946 he recommended the planned procurement of 100 B-36 aircraft be reduced to just a few test aircraft. He feared the plane's performance, such as speed and range, and other characteristics were not good enough at that point for SAC and the nation to pin their hopes on it. He preferred the Boeing B-50 as SAC's new atomic bomb carrier.[17]

Many AAF leaders, including SAC's commander, preferred the Boeing B-50 as SAC's new atomic bomb carrier. (U.S. Air Force)

But others in high places, including Lieutenant General Nathan Twining, then Commanding General of Air Materiel Command (AMC) and later Chairman of the Joint Chiefs of Staff (JCS), disagreed. Twining believed it was too early to judge the aircraft with the XB-36 just entering testing. He expected improvements based on the testing, and he reasoned that the B-36 was the only aircraft far enough along to fill the atomic bomb carrier role.

During April through June 1948, the Air Force conducted an evaluation to compare B-36 range, speed, altitude, and bombload capacity against the B-50. The United States was developing a strategy of deterrence to counter an atomic attack from the Soviet Union. They needed a long-range bomber capable of carrying the bomb to hit targets in Soviet territory. Only the B-36 could complete this mission without aerial refueling. The results of the evaluation thus favored the B-36. On 18 June, the Berlin Blockade began and spared the Air Force the decision to cancel the B-36. Now was not the time to cancel weapons and signal a weakness or lack of commitment to use atomic weapons.

On 25 June 1948, the Secretary of the Air Force, W. Stuart Symington, and other Air Force officials unanimously agreed to continue the B-36 program. Twining continued to argue that the B-36 was the best plane for its purpose available and would be improved with numerous planned modifications and upgrades. General Carl A. "Tooey" Spaatz,

Air Force Chief of Staff, agreed. Although the aircraft had some shortcomings, the group agreed they could be overcome. The commencement of the Berlin Blockade increased their concern about Soviet aggression. The B-36 program continued in full swing. General Kenney, now convinced, returned to his earlier position as one of its enthusiastic supporters.[18]

By early 1949 General Vanderburg's vision of a 70-group Air Force was quickly fading. President Truman's decision to hold the 1949 defense budget to $11 billion meant the Air Force had to reduce from the 55 groups it had built to only 48 groups. Once again, the B-36 was under scrutiny as the Air Force cancelled the purchase of various bombers, fighters, and transport aircraft in January 1949. But Lieutenant General Curtis LeMay, SAC Commanding General since October 1948, recommended procurement of additional B-36 aircraft for both the bombing and reconnaissance roles. He believed the B-36 could do everything as well or better than the proposed B-54 bomber that could be cancelled instead. He anticipated increasing the force by two heavy B-36 bomb groups and one RB-36 strategic reconnaissance group at the expense of two medium groups. Each group size would increase from 18 Unit Equipped (UE) aircraft to 30 UE which would reduce personnel cost and improve combat power. The president authorized the release of funds for the procurement of 39 additional B-36 bombers on 8 April 1949 and procurement of the RB-36 aircraft on 4 May 1949.

President Truman announced that the Soviet Union had successfully exploded its first atomic bomb on 23 September 1949. On 31 January 1950, President Truman directed the Atomic Energy Commission (AEC) to continue all work on all forms of "atomic energy" weapons to include the hydrogen bomb (also called the super bomb). The B-36, the greatest aerial delivery system ever produced in the world up to that time, thus approached its rendezvous with the Cold War. As a top SAC officer put it, the postwar Air Force was faced with the need to "develop a powerful strategic force in being, able to deliver such massive retaliation that any aggressor nation would hesitate to launch an attack against our homeland. Part of the fleet had to be able to launch additional sizable strikes from our North American bases. The B-36 gave us this capability."[19]

Roles and Missions

Strategic Bombardment

WWII proved the value of both Naval and Army aviation. Navy carriers proved their ability to operate in remote regions of the world and dominate the skies. Their ability to attack land targets was invaluable in allowing the Marines to capture remote Pacific islands putting B-29 bombers within reach of Japan. The AAF proved its ability to conduct strategic bombing campaigns in both Europe and Japan and destroy the enemy's war making capability. General Spaatz, the newly appointed Commanding General of the AAF, capitalized on this experience from WWII and formed three new combat commands on 21 March 1946 including SAC, Tactical Air Command (TAC), and Air Defense Command (ADC).

Spaatz defined SAC's mission as: [SAC] will be prepared to conduct long range offensive operations in any part of the world either independently or in cooperation with land and Naval forces; to conduct maximum range reconnaissance over land or sea either independently or in cooperation with land and Naval forces; to provide combat units capable of intense or sustained combat operations employing the latest and most advanced weapons; to train units and personnel for the maintenance of the Strategic Forces in all parts of the world; to perform such special missions as the Commanding General, Army Air Forces may direct.[1]

Congress created the independent Air Force on 18 September 1947 with a planned fleet twice the size of the Navy's – 24,000 aircraft to the Navy's 11,500 – and only the Air Force would have heavy bombers. These bombers would be under the control of SAC. With the emergence of Joseph Stalin's ambitions, SAC's strategic mission quickly rose in importance and there was no longer any question who the "enemy" was. By happenstance, the long-distance payload of the B-36 equaled the weight of one atomic bomb – roughly 10,000 pounds – and its combat radius equaled the great-circle route from Maine to Leningrad.[2]

By 1949, the Air Force was fully committed to strategic bombing based on the successes of WWII and the philosophies of pre-war airpower advocates like Giulio Douhet, Billy Mitchell, and Hugh Trenchard. Air Force leaders fully accepted that the "battleplane" was the key to airpower and "auxiliary airpower" supporting ground operations was worthless. They believed success in future wars was dependent on massive air offenses to crush critical enemy industrial, economic, and military resources and their will to fight. This could only be done through an independent Air Force with massive amounts of battleplanes focused on strategic bombing operations. These battleplanes should have the range, speed, and self-defense needed to reach and bomb enemy targets. The B-36 armed with atomic weapons was the battleplane the Air Force envisioned for this role.

Air Force leaders envisioned an independent Air Force with massive amounts of "battleplanes" focused on strategic bombing operations with the range, speed, and self-defense needed to reach and bomb enemy targets. (American Aviation Historical Society)

Pending the arrival of its new $5.7-million-dollar baby, SAC made do with 160 veteran B-29 Superfortress aircraft, and it was these aircraft that answered the call to deploy to European bases when the Russians shut off ground access to Berlin in the summer of 1948. It was a colossal bluff. In all of SAC, only 27 B-29 aircraft had the "Silver Plate" modifications needed to carry an atomic bomb, and these were all assigned to the 509th Bomb Group (BG), which stayed home. As for bombs, the United States "stockpile" contained exactly 13, controlled by the AEC, and President Harry Truman refused to say if he would ever release them to the military. Even if he had given the order to launch an attack, the 509th would have needed five days to pack up, fly to an AEC depot, load the nukes, and move overseas.[3]

Revolt of the Admirals
In 1949 a battle raged between the Navy and the newly formed Air Force over roles and missions which became known as the Revolt of the Admirals. After the National Defense

Act was signed in 1947 the Air Force was established as a separate service. Much to the disappointment of AAF leaders, such as General Arnold and Lieutenant General Ira C. Eaker, AAF Deputy and Chief of Staff, the act was a compromise that legitimized four military air forces.[4] Despite this, Navy leaders were concerned that creation of the Air Force would soon mean the loss of their own aviation role and assets to the Air Force. Furthermore, limited, and equal defense budgets after WWII meant that each service struggled to build their forces. The Air Force had a goal of 70 groups and was only able to achieve 55, and this was later reduced to 48.

Meanwhile, Secretary of the Navy John L. Sullivan announced in early February 1948 that the Navy planned to build a flush-deck supercarrier. Although Sullivan insisted the Navy had no intention of usurping the strategic mission, Secretary of the Air Force Stuart Symington and General Spaatz, who was now Air Force Chief of Staff, thought that the Navy was in the process of building a strategic air force with the planned supercarrier and its long-range patrol bombers at its core.[5]

During a March 1948 meeting in Key West, FL the Joint Chiefs of Staff (JCS) agreed that Strategic Bombing and build-up of an atomic bomb delivery system was the key to defense against Soviet aggression. The JCS also agreed that the Air Force was responsible for strategic bombing operations and not the Navy. The Air Force proposed the B-36 bomber as the core of this force. However, the Navy still proposed a supercarrier capable of launching nuclear armed aircraft. They complained that the B-36 was incapable of performing the mission.

Top officials of the National Military Establishment and the JCS met with Secretary of Defense James Forrestal in Key West, FL in March 1948. (Harry S. Truman Presidential Library and Museum)

The second roles-and-missions conference, held at Newport, RI in August 1948, saw Secretary of Defense James Forrestal and the Joint Chiefs in agreement that the Air Force would have primary responsibility for strategic bombing, but during war would be supplemented by Naval forces. The agreement stated that "the service having the primary function must determine the requirements but must take into account the contributions which may be made by forces from other services." The JCS also decided that the Chief of the Armed Forces Special Weapons Project would report to General Hoyt S. Vandenberg, who had succeeded Spaatz as Air Force Chief of Staff, in effect giving the Air Force operational control of the atomic bomb, something it had long desired.[6]

The Navy still insisted that it was more capable of performing the strategic nuclear delivery role with its supercarrier and nuclear capable bombers. However, this was an expensive proposition and with limited defense budgets and equal shares going to each service the supercarrier was viewed within the defense department as unworkable. The Air Force pressed ahead in the building of its atomic deterrent force. The Berlin blockade had stunned the world in 1948, and in October, General Vandenberg directed Lieutenant General Curtis E. LeMay to head SAC, replacing General George C. Kenney. Vandenberg, with the backing of Symington and LeMay, supported production of the B-36 long-range strategic bomber. In December 1948, Vandenberg and top Air Force

Artist's conception of the U.S. Navy aircraft carrier USS United States (CVA-58) by Bruno Figallo, October 1948, showing the ship's approximate planned configuration. Locations for smokestacks, elevators, and bridge not yet decided. (Naval History and Heritage Command)

commanders met at Maxwell AFB, AL and decided that the structuring of SAC's atomic force should be their highest priority.[7]

Louis Johnson replaced Forrestal as Secretary of Defense in March 1949. Johnson was a wealthy lawyer, aspiring Politian, and former official with Convair. That last connection, which today would seem like a scandal worthy of a special prosecutor, was common at the time. Who knew more about weapons than the men who built them?[8] Johnson believed that construction of the flush-deck supercarrier was unnecessary, wasteful of funds, and a duplication of the Air Force's mission. When Truman ordered him to reduce costs and economize, Johnson polled the Joint Chiefs (Admiral Louis E. Denfeld, Chief of Naval Operations, was the lone vote for construction) and then obtained Truman's approval to stop construction of the carrier on 23 April 1949.[9]

Johnson exercised the newly granted authority of the Secretary of Defense from the amended National Security Act of 1949 and made the unilateral decision without consultation of either the Secretary of the Navy or the Chief of Naval Operations. An irate Secretary of the Navy Sullivan immediately resigned. He wrote Johnson that this action "represented the first attempt ever made in this country to prevent the development of a power weapon. The conviction that this will result in a renewed effort to abolish the Marine Corps and to transfer all Naval and Marine aviation elsewhere adds to my anxiety." The battle had been joined.[10]

In April to May 1949 an anonymous letter began to circulate in Washington that claimed the B-36 was a "billion-dollar blunder" and "lumbering cow" and questioned the B-36 operational capabilities. It condemned the B-36 as "an obsolete and unsuccessful aircraft" and charged that the Air Force had acquired it only after Convair had contributed $6.5 million to various Democratic politicians.[11] It also intimated that Secretary Johnson owed favors to Convair and thus supported the bomber's construction.

The Navy also claimed it had at least three jet fighters that could leave the monster behind at 40,000 feet. The admirals wanted a matchup, but they would never get one. The JCS told Johnson the test was a bad idea, and the Air Force said it had already demonstrated that fighters could not maneuver at that altitude. In the Air Force demonstration, simulated B-36 attacks on bases in Florida and California were met by three front-line fighters: a North American F-86A Sabre, a Lockheed F-80C Shooting Star, and a Republic F-84 Thunderjet. Radar picked up the intruder 30 minutes out; the fighters took 26 minutes to climb to 40,000 feet and another two minutes to find the B-36. The fighters were faster than the big bomber, but their wing loading was so high that they could not turn with the bomber without stalling in the thin air. Even if a B-36 were detected and Soviet fighters caught it, the pilot could evade them by making S-turns, said the Air Force.[12]

Although navy fighters were faster than the huge B-36 bomber, the Air Force claimed the bomber pilot could evade them by simply making S turns since the fighter wing loading was so high that they could not turn with the bomber without stalling in the thin air at high altitudes. (American Aviation Historical Society)

Congressman James Van Zandt, Republican from Pennsylvania, and Captain in the Naval Reserve, called for a special panel to investigate the charges including a comprehensive investigation of the B-36 and the decision to cancel the supercarrier. These charges and the attendant congressional hearings received national attention and a great deal of coverage in the press.

During 9-25 August 1949, the House Armed Services Committee deliberated over the B-36. With Chairman Carl Vinson of Georgia presiding, it was soon revealed that the author of the anonymous letter was Cedric Worth, a civilian assistant to Under Secretary of the Navy Dan Kimball. He was aided with input from Commander Thomas Davies, assistant head of Op-23, a Navy research group formed to gather material critical of the B-36. He was also aided by Glenn Martin whose company lost a contract in favor of the B-36.

Under the pressure of the hearing, Worth was forced to recant everything and admit that he made a grave error. Consequently, the House committee found no evidence to substantiate the charges and cleared all Air Force officials. There was "not one iota, not one scintilla of evidence," emphasized Vinson, "that would support charges that collusion, fraud, corruption, influence, or favoritism played any part whatsoever in the procurement of the B-36 bomber."[13]

In October, twelve days of spectacular hearings were conducted on "Unification and Strategy". The Navy refuted the Air Force position in a mostly emotional appeal and claimed that the B-36 was an inferior plane that could not accomplish the strategic bombing mission against the Soviet Union, that the entire concept of strategic bombing was unsound, and that the decision not to construct the supercarrier weakened the Navy and was itself a threat to the national security. They believed the carrier gave significant flexibility for "strike" missions such as attacking convoys, supporting amphibious operations, bombing coastal targets, and even strategic bombing of inland targets. Many of the Navy's arguments had been validated in the Strategic Bombing Survey commissioned on 3 November 1944 by President Roosevelt and a similar study ordered by President Truman on 15 August 1945.

The North American Aviation AJ-1 Savage was a nuclear bomber designed to carry atomic weapons from aircraft carriers. (U.S. Navy)

Vice Admiral Arthur W. Radford, Captain Arleigh Burke, and Admiral Denfeld testified for the Navy. Radford termed the B-36 a "bad gamble" and indicted what he called the "atomic blitz", the land-based strategic deterrent. He also criticized the lack of attention the Air Force had paid to tactical fighter and attack aircraft. Burke argued that the United States is a maritime nation and the capability to project power from the sea is critical for national defense. This required a strong Navy. He trumpeted that carrier aviation was more versatile than land-based airpower and could be enhanced with the ability to carry nuclear capable bombers as proposed for the supercarrier. Denfeld declared himself "gravely concerned" about the Navy's ability to carry out its mission without such a weapon as the supercarrier.[14]

While the Navy's presentation was disjointed and emotional, the Air Force presented a more coherent and polished case. The Air Force argued that the B-36 carrying atomic bombs could achieve victory without large United States manpower losses. They said the B-36 could do the job for half the cost of any other weapon. They believed they could purchase 500 B-36 aircraft for the cost of one supercarrier and expose 242 times fewer men to danger. Their overall position was that airpower was now the dominant military force, that the Air Force was the rightful component of airpower, and strategic bombing was the most important function of the Air Force.

Secretary Symington and General Vandenberg were the primary witnesses for the Air Force. Symington refuted the B-36 charges and emphasized that the concept of strategic bombing had been approved by the Joint Chiefs and assigned to the Air Force. The attacks against the Air Force, declared Symington, "imperiled the security of the United States. It was bad enough to have given a possible aggressor technical and operating details of our newest and latest equipment... It is far worse to have opened up to him in such detail the military doctrines of how this country would be defended." The B-36 intercontinental bomber, noted Symington, was under attack by Naval officials because it was a threat to the Navy.[15]

Vandenberg stressed the effectiveness of strategic bombing and confirmed the B-36 could do its mission. He also stated that he could see no strategic mission for the supercarrier. General Arnold was called out of retirement to testify. In a powerful and sharply worded testimony he chastised the Committee and the detractors of the B-36 for attempting to disrupt the development of an immediate strategic deterrence requirement and for giving away secret performance information in open hearings. He praised Kenney and LeMay for taking the responsibility to build a believable strategic force while others whimpered about what ought to be done.[16]

Army General Omar Bradley, Chairman of the JCS, also testified. He confirmed that the JCS had assigned strategic bombing responsibility to the Air Force and given it top priority. He supported the B-36 and believed the Navy was upset over the cancellation of the supercarrier and in open rebellion against civilian control of the military. He criticized the Navy's methods and accused its officers of poor leadership, disloyalty, and failure to accept unity of command.

In the end, Congress agreed with the Air Force and allowed the B-36 bomber production to continue. The Navy lost its supercarrier and claim to the strategic nuclear delivery mission... at least for a time. The advent of Polaris missiles fired from nuclear submarines eventually made the Navy the third leg of the strategic triad along with Air Force bombers and land-based missiles.

Strategic Reconnaissance

In the late 1940s, strategic intelligence on Soviet capabilities and intentions was scarce. When Lieutenant General LeMay took command of SAC in 1948 he insisted, based on his experience with B-29 strikes during WWII, that SAC needed a strategic reconnaissance capability. In September 1948, the Air Force approved the urgent conversion of B-36A aircraft to the RB-36E configuration as an immediate response. During the Board

The last B-36A (44-92014) to be converted to an RB-36E sits on the ramp at Convair Fort Worth 17 April 1951. (University of North Texas Libraries)

of Senior Officers meeting in November, Air Force senior leaders selected the RB-36D along with the RB-47 for additional reconnaissance capability to supplement the RB-36E. Deliveries to SAC began in June 1950, but the RB-36D fleet did not become operationally capable until June 1951 due to material shortages.

The RB-36D aircraft was easily recognizable by the bright aluminum skin covering the camera compartment. (U.S. Air Force)

One of SAC's initial missions was to plan strategic aerial reconnaissance on a global scale. The first efforts were in photoreconnaissance and mapping. Along with the photo-reconnaissance mission, a small Electronic Intelligence (ELINT) cadre was operating. Weather reconnaissance was also part of the effort, as was Long Range Detection – the search for Soviet atomic explosions. Before the development of the Lockheed U-2 high altitude spy plane and Corona orbital reconnaissance satellites, technology and politics limited American reconnaissance efforts to the borders, and not the heartland, of the Soviet Union.[17]

With a range of 9,300 miles, RB-36D aircraft began probing the boundaries of the Soviet Arctic in 1951. Although onboard equipment indicated detection by Soviet radar, interceptions at the B-36's service ceiling would have remained difficult. RB-36 aircraft operating from Royal Air Force (RAF) Sculthorpe in England made several overflights of

Soviet Arctic bases, particularly the new nuclear weapons test complex at Novaya Zemlya. RB-36 aircraft performed several rarely acknowledged reconnaissance missions and are believed to have frequently penetrated Chinese (and Soviet) airspace. However, advances in Soviet air defense systems meant that the RB-36 became limited to flying outside the borders of the Soviet Union, as well as Eastern Europe.[18]

In late 1952, during the Korean War, six RB-36D aircraft from the 5th Strategic Reconnaissance Wing (SRW) at Travis AFB, CA were deployed to the 91st Strategic Reconnaissance Group (SRG) at Yokota AB, Japan. This was the first introduction of RB-36 aircraft to the Korean theater. While not employed in any combat missions over North Korea, these RB-36s conducted high altitude aerial reconnaissance over Chinese Manchurian and Soviet East Asian targets while attached to the 91st SRG.[19]

RB-36D aircraft conducted high altitude aerial reconnaissance over Chinese Manchurian and Soviet East Asian targets during the Korean War. (U.S. Air Force)

Models and Variants

The B-36 was a very versatile aircraft due to its large size, range, and high-altitude capability. In its day, it was a relative monster. The original designs included two configurations, one with four engines and one with six engines, both using "pusher" props. The propellers were 19 feet in diameter, and to keep the tips from going supersonic they were geared to turn less than half as fast as the engines. The engines and propellers produced an unforgettable throbbing sound when the B-36 flew overhead.[1] Both configurations also featured a twin tail that was later replaced with a single tail.

The B-36 was the largest piston engine bomber ever built and, at 230 feet, still holds the record for the longest wingspan of any American combat aircraft. It eclipsed the other aircraft of its day such as the B-29. As shown in the following figure, there were many models and variants developed during its 10-year service life. They were ever evolving, and the fleet was in a constant state of flux as various configurations were developed and implemented. For example, all B-36A models were converted to RB-36E reconnaissance models, and most B-36B models were converted to B-36D with the addition of four J47 engines on the outer wings.

Crewmembers could move between forward and aft compartments through an 85-foot-long pressurized tube. (American Aviation Historical Society)

The aircraft had both a forward and aft crew compartment. The aft compartment supported gunnery stations and contained bunks for crew rest. Crew members could transit between the two compartments using a small flat cart about 5-feet long with four wheels that rode in tracks by pulling themselves on an overhead cable through an 85-foot long, 36-inch diameter, pressurized tube (though they rarely did as most preferred catnaps on the floor of the radio room in the main cabin, with what little time they had for sleeping). The wing roots were 7.5 feet thick with catwalks that allowed crewmen (usually gunners) access to manually reset the engine circuit breakers and landing gear down-locks if required during flight. Each airplane had 336 spark plugs and after most flights, which typically lasted a day and a half, a mechanic would have to haul a bucket of replacement plugs to the airplane to service all six engines, a process that took several hours to complete. The standard plugs were

B-36 PEACEMAKER: THE BIG STICK OF STRATEGIC AIR COMMAND

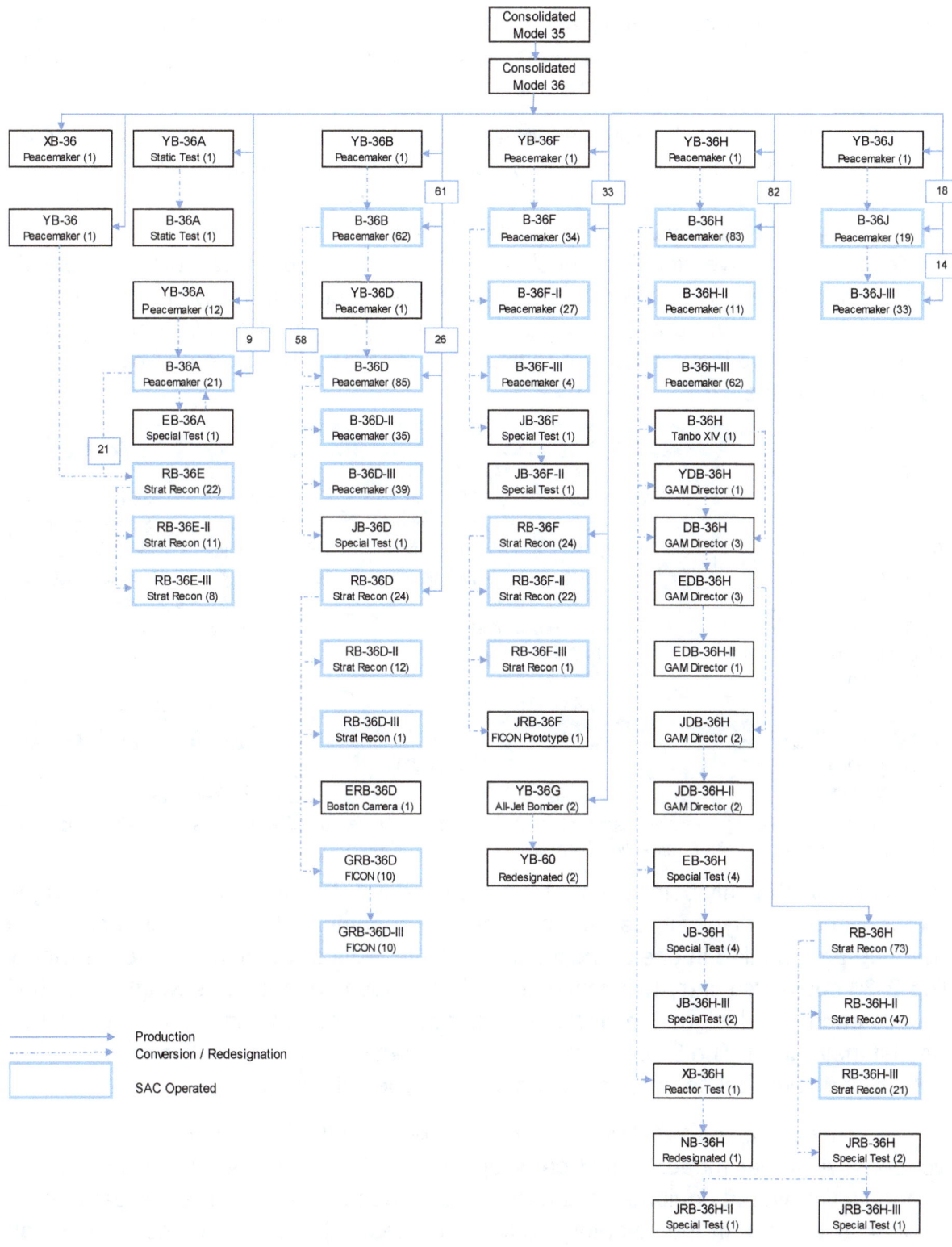

The versatility of the B-36 design was apparent in its numerous models and variants.

later replaced with longer lasting platinum plugs, which significantly reduced the maintenance frequency. The engines leaked oil, and sometimes the Flight Engineer (FE) had to shut one down in flight because it had exhausted its allotment of 150 gallons.

The B-36 surpassed all Air Force expectations as a bomb delivery platform. It carried the largest internal bombload, at 86,000 pounds, of any American bomber and is only surpassed worldwide by the Russian Tu-160 Blackjack bomber, which can carry 88,000 pounds in two internal bays. With its four bomb bays reconfigured, the B-36 could carry a wide variety of conventional and nuclear bombs as depicted by the typical bombloads shown below. The lower portion of the bulkhead between Bomb Bay No. 1 and No. 2 swung aside to allow the aircraft to carry 12,000, 22,000, and 43,000-pound bombs, which occupied two bomb bays due to their immense size. The bulkhead between No. 3 and No. 4 did the same.

Conventional	Bomb Bay No. 1	Bomb Bay No. 2	Bomb Bay No. 3	Bomb Bay No. 4
10,000 pounds		14x 500 pounds	6x 500 pounds	
48,000 pounds	3x 4,000 pounds	3x 4,000 pounds	3x 4,000 pounds	3x 4,000 pounds
56,000 pounds	8x 2,000 pounds	6x 2,000 pounds	6x 2,000 pounds	8x 2,000 pounds
66,000 pounds	38x 500 pounds	28x 500 pounds	28x 500 pounds	38x 500 pounds
72,000 pounds	20x 1,000 pounds	16x 1,000 pounds	16x 1,000 pounds	20x 1,000 pounds
48,000 pounds	2x 12,000 pounds		2x 12,000 pounds	
65,000 pounds	1x 22,000 pounds		1x 43,000 pounds	
66,000 pounds	1x 22,000 pounds		2x 22,000 pounds	
67,000 pounds	2x 12,000 pounds		1x 43,000 pounds	
86,000 pounds	1x 43,000 pounds		1x 43,000 pounds	
Nuclear	**Bomb Bay No. 1**	**Bomb Bay No. 2**	**Bomb Bay No. 3**	**Bomb Bay No. 4**
50,500 pounds		1x 8,500 pounds	1x 42,000 pounds	
84,000 pounds	1 x 42,000 pounds		1 x 42,000 pounds	

The B-36 was the only bomber capable of carrying two 42,000-pound Mk-17 thermonuclear weapons or two 43,000-pound T-12 Cloudmaker "earthquake effect" conventional bombs.

The B-36 was the only bomber capable of carrying two 42,000-pound Mk-17 hydrogen bombs, although it typically carried only one in the (combined) two aft bomb bays with an 8,500-pound Mk-6 in one of the forward bays to reduce weight and increase range. The B-36 could also carry two conventional T-12 Cloudmaker bombs weighing 43,000 pounds each and designed to create an "earthquake effect" to shake the ground when they hit their target. The B-36 could carry 67 different types of conventional, incendiary, cluster, and chemical bombs, as well as several types of mines.

The B-36A was limited to 72,000 pounds total bombload and could not carry nuclear weapons. All other models could carry up to 86,000 pounds total including nuclear weapons. On aircraft equipped to carry nuclear weapons, any airborne nuclear or thermonuclear weapon in the inventory could be carried – the B-36 was the only aircraft that could do so. Only a single type of bomb could be carried within each bomb bay, although each of the four bomb bays could carry different types of bombs if necessary.

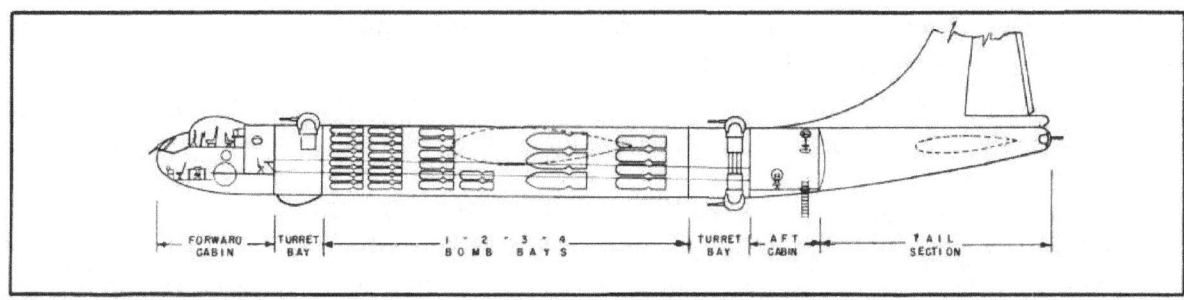

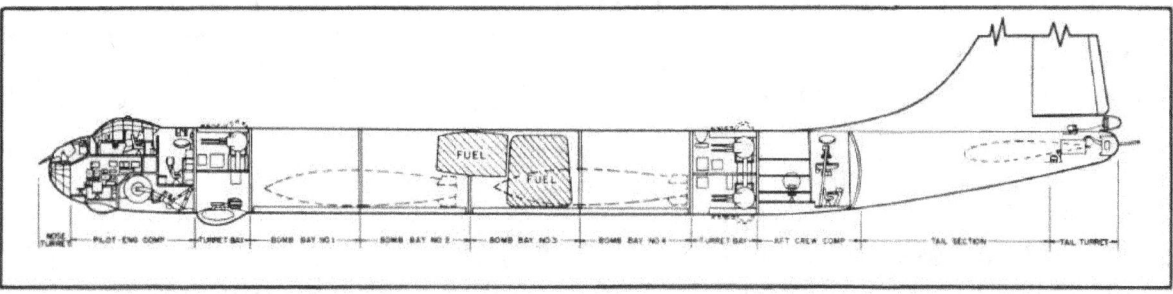

The B-36 interior was comprised of enormous bomb bays capable of carrying up to 86,000 pounds of bombs of various sizes including 500-, 1,000-, 2000-, 4,000-pound, and 43,000-pound T-12 Cloudmaker conventional bombs and the 42,000-pound Mk-17 thermonuclear bomb. (U.S. Air Force)

Although it was universally considered a nuclear bomber, like all other aircraft designed during WWII the B-36 was not originally designed as one. After the war, the B-36 was gradually modified with nuclear capability and assumed the nuclear delivery role. By the end of 1947, the Air Force had only 32 B-29 aircraft modified under Project SADDLE TREE[2] available to carry atomic weapons and they were all assigned to the 509th BG. Between May 1947 and June 1948, 18 B-36B aircraft were modified under Project SADDLE TREE and the remaining B-models were scheduled to come off the production line with most of the changes already in place.[3]

By the end of 1950, SAC had 52 B-36 aircraft equipped with the SADDLE TREE capability. These aircraft also incorporated Project GLOBAL ELECTRONICS MODIFICATION (GEM) that allowed the bombers to navigate over and around the Arctic Circle. At the same time the Air Force initiated Project ON TOP that allowed B-36 aircraft to carry Mk-4, Mk-5, and Mk-6 atomic devices. It also began the Universal Bomb Suspension (UBS) system development that allowed the B-36 and other Air Force bombers to be easily configured to carry most nuclear weapons. The UBS allowed the B-36 to carry Mk-4, Mk-5, Mk-6, Mk-8, and Mk-18 atomic bombs and Mk-15, Mk-21, Mk-36, and Mk-39 thermonuclear weapons as a single nuclear weapon in Bomb Bay No. 1.[4]

In July 1950 SAC decided that the B-36 should be able to carry more than one atomic bomb at a time. At least 30 aircraft (12 B-36D and 18 B-36H) were subsequently modified to carry the UBS in all four bomb bays. Some other aircraft were modified to carry the UBS in Bomb Bay No.1 and No. 4 only. The factory began to equip aircraft with the UBS beginning with the B-36F and all B-36H aircraft were equipped with the UBS in Bomb Bay No. 1 and No. 4. Some aircraft were subsequently upgraded to carry nuclear weapons in Bomb Bay No. 2 and No. 3 also.

By the end of 1953, 20 B-36 aircraft were modified to carry the Mk-14 and Mk-17 thermonuclear devices. By the middle of 1955, 208 aircraft (which accounted for most of the B-36 bomber variants in SAC inventory) could carry these 20-megaton devices – the most powerful weapon ever deployed and only deliverable by the B-36.[5]

Beginning in 1952 all RB-36 aircraft were modified to carry nuclear weapons in Bomb Bay No. 4. When SAC began converting its RB-36 fleet to a dual bomber/reconnaissance role in 1954 all RB-36 aircraft had the UBS installed in Bomb Bay No. 2 and some had provisions to carry nuclear weapons in Bomb Bay No. 3.

An auxiliary fuel tank with 3,000-gallon capacity could be carried in Bomb Bay No. 3 on all aircraft. Another could be installed in Bomb Bay No. 2 on some aircraft. The B-36B could be field modified to carry up to four bomb bay fuel tanks as shown, but the 328,000-pound gross weight only allowed for three bomb bays carrying a maximum of 7,393 gallons of bomb bay fuel. Presumably, all other models could carry these four bomb bay tanks, since their gross weight exceeded 328,000 pounds, but similar depictions show only the No. 3 bomb bay tank installed. The auxiliary fuel tanks could be dropped in flight, but the fuel and power connections had to be manually disconnected.[6]

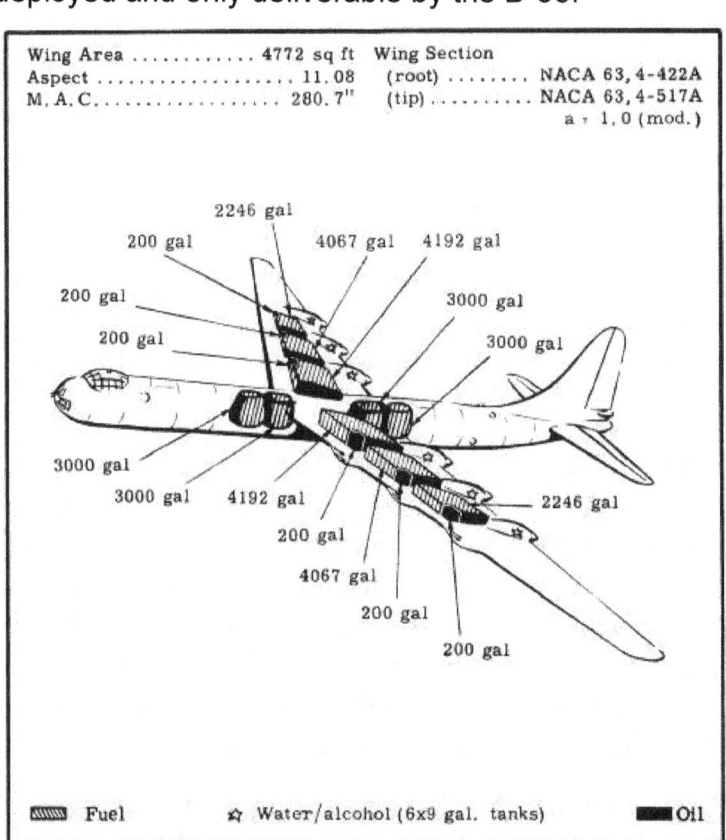

The B-36B was designed to carry up to four auxiliary bomb bay tanks adding 12,000 gallons of available fuel. (U.S. Air Force)

Convair supplied cargo carriers with each aircraft designed to aid in aircraft deployments. Two cargo carriers could be uploaded into Bomb bays No. 1 and No. 4, and one

Mk-17 Thermonuclear Bomb

The Mk-17 was the largest thermonuclear bomb ever developed by the United States. The B-36 was the only bomber with the size and capability to carry two of these devices, when properly configured, although it typically carried only one in the (combined) two aft bays and an 8,500-pound Mk-6 in the one of the forward bays to reduce weight and increase range.

The Mk-17 was tested as the TX-17 from November 1952 to December 1953. The thermonuclear version was tested in March and April of 1954. The tests yielded 11 to 13.5 megatons, but the weapon was ultimately classified as a 20-megaton device. The bomb was authorized to be stockpiled in April 1954 under Emergency Capability 17 (EC-17), becoming the first thermonuclear device to be stockpiled under a regular production program. Approximately 200 were produced.

The Mk-17 was the largest air dropped thermonuclear bomb ever developed by the United States. (U.S. Air Force)

The bomb was huge weighing in at 41,400 to 42,000 pounds. It was 24.7 feet long, 61.4 inches in diameter, and the casing was 3.5 inches thick. The aircraft would suddenly soar hundreds of feet upward when the bomb was dropped, while the pilot fought to regain control. Drop tests were conducted from November 1954 to late 1955 to determine the best parachute configuration that would allow the aircraft to escape the blast after release. The final design used a 5-foot guide chute, a 16-foot deployment chute, and a 64-foot main canopy released in stages.

The bomb could be dropped safely from a B-36 at 40,000 feet using this parachute configuration. However, lower altitudes would cause high shock loadings. Also, the high speed of the B-52, seen as a superior capability over the B-36 by Air Force leaders, caused high shock loadings and prevented the aircraft from dropping the Mk-17. Further parachute testing was accomplished to allow the B-52 to drop the bomb and the first phase of testing, completed on 19 January 1955, indicated that a solution was possible. Ultimately, the requirement for carriage of the bomb on the B-52 was cancelled in April 1956 due to release system problems.

The bomb was phased out beginning in October 1956 and was completed by October 1957. The Air Force converted to the Mk-36, which was now entering service in large enough quantities and provided a higher yield with less weight. This bomb could be carried by all SAC aircraft including the B-47 and B-52. A total of 920 were produced.

in No. 2 and No.3. These cargo carriers allowed the B-36 to carry along its own tools and equipment during deployments.

These carriers were put to good use on many occasions. During a typical deployment to Guam, for example, light weight vehicles and other equipment were hoisted into the bomb bays and transported with their aircraft and crews. B-36 aircraft typically used these cargo carriers to transport the equipment required to upload nuclear weapons into the aircraft. When they landed at the designated weapons storage airfield, they swapped the cargo carriers for weapons before deploying to their forward operating

B-36 spare engine pods capable of carrying two R-4360 engines and accessories each. (U.S. Air Force)

location. Spare engine pods, each capable of carrying two R-4360 engines and accessories, were also available. These only worked with the original sliding bomb bay doors and were never used operationally.

Like all new aircraft the early models had significant problems. The supposedly improved, higher-horsepower engine was disappointing. Mounting engines on the trailing edge of the wing created unexpected vibration difficulties. Wing metal fatigue was the result. Skin panels loosened with the stress of flight causing the fuel tanks to leak, sometimes severely. New materials and fabrication methods had to be developed including a specially developed fuel tank sealant. Sealing the tanks was a complicated process that took hours to complete for each tank but was eventually mostly successful in preventing leaks.

By 1951, the aircraft were nearly combat ready, but many problems remained. Chief among them was the defensive armament system, which SAC viewed as the weakest link in the aircraft's capability. The Air Force conducted a series of tests from April 1952 to mid-1953 and concluded that system improvements, along with improved maintenance and gunnery crew training, were needed. However, the systems were never fully operational or effective. All defensive systems except the tail gun and its radar were eventually removed as part of the Featherweight III modifications designed to reduce aircraft weight and increase range.

Prototypes

XB-36. The XB-36 (42-13750) was the fruition of the original specifications issued in 1941, and the Consolidated Aircraft proposed design. Although it was a very successful proof of concept, Consolidated Aircraft engineers struggled with several design issues and priority changes that threatened the program. For example, during initial wind tunnel tests conducted at the Massachusetts Institute of Technology (MIT) and the California Institute of Technology (CALTECH) in July 1942, drag and weight was higher than expected. An Air Force committee inspected the XB-36 mock-up on 20 July 1942. They were concerned about the drag and weight issues, which needed to be reduced to meet the 10,000-mile range requirement. Some members of the committee wanted to eliminate the guns to reduce weight, but others argued this would make the aircraft essentially useless. This debate nearly caused cancellation of the program, but the committee ultimately agreed to remove other "less needed" items. Although a new model was constructed to address drag and weight reduction issues, it could not be retested for over a year due to war priorities.[7] Development continued in the meantime.

Consolidated Aircraft and the Air Force continued to tackle XB-36 technical problems for the next three years. Weight management was a constant problem. Consolidated Aircraft was concerned with engine performance versus aircraft weight, but even the new X-WASP engines provided by P&W added 2,304 pounds. Adding a nose gun turret caused redesign of the nose section. Radio and radar equipment increased weight by at least 3,500 pounds, and more if the AN/APQ-7 Eagle radar bombing system antenna could not be integrated into the wing leading edge as planned.[8]

Consolidated Aircraft, in conjunction with Wright Field, also developed a new 3,000 pounds per square inch (psi) hydraulic system to move the huge landing gear and flaps required for the B-36. It replaced the 1,500 psi systems standard at the time (3,000 psi then became the standard for all aircraft until Bell developed the 5,000-psi system for the V-22). Wright Field had been developing a 400 cycle, 208 volt, 3-phase Alternating Current (AC) system to supplement the Direct Current (DC) systems used on other aircraft and Consolidated Aircraft used it on the B-36. Combined AC and DC systems became standard on aircraft after this innovative approach was proven on the B-36. Finally, Consolidated Aircraft replaced the usual carbon-dioxide fire extinguishers with

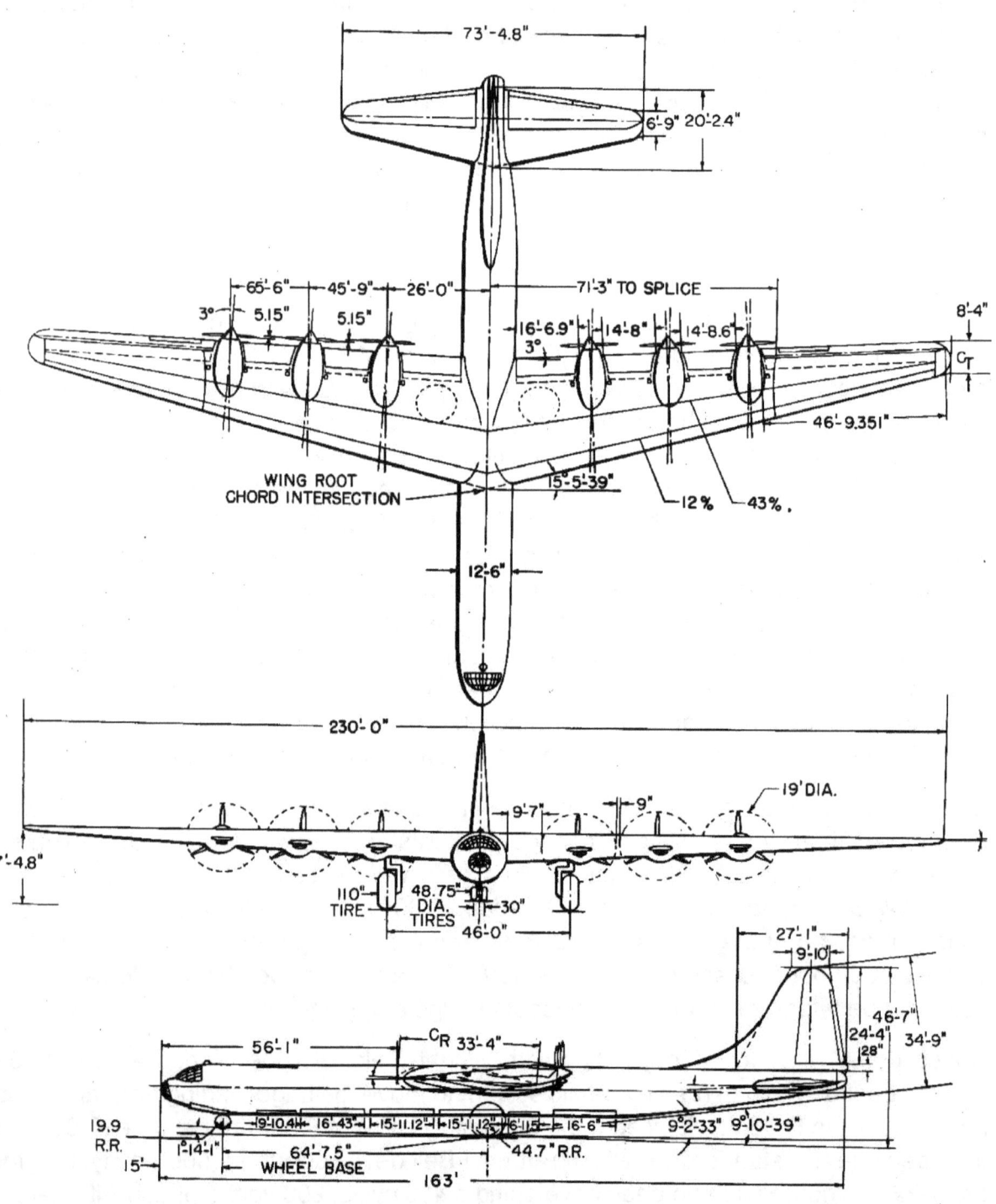

The XB-36 introduced many technical advancements, but weight management was a constant concern and drastically affected the aircraft's range predictions. (American Aviation Historical Society)

methyl-bromide because it had a lower boiling point and could be stored at lower pressures in lighter cylinders.[9]

Much of the aircraft was covered with magnesium skin to reduce weight, except the forward and aft pressurized compartments which were covered with aluminum. This resulted in a distinctive contrast between the bright aluminum and dull magnesium skin sections. Consolidated Aircraft developed a metal adhesive, in both liquid and tape forms, to attach exterior skin and eliminate welding and pop rivets (welding allowed 12 million cycles, rivets 18 million, and adhesive bonding 240 million cycles). This adhesive was meant to reduce production time, and it proved useful on the B-36 magnesium skin but was not as effective on other applications. Despite its promise, crews sometimes arrived at the aircraft greeted with dozens of sheetmetal mechanics working on loosening upper wing skins that had to be reattached before flight.

The XB-36 aircraft rolled out of the Convair plant in Fort Worth on 8 September 1945. After several promised first flight dates came and went, the aircraft finally flew on 8

The XB-36 (42-13750) completed its first flight on 8 August 1946 just days before Japan's surrender from WWII. (U.S. Air Force)

August 1946 and completed a 37-minute flight just days before Japan's surrender from WWII. This flight was considered generally successful. The aircraft flew several more test flights that determined the aircraft's top cruising speed was 230 mph and average was 216 mph. Maximum speed was 345 mph at 35,000 feet. Two major problems were soon discovered – a lack of proper engine cooling and a propeller vibration. A two-speed cooling fan was eventually developed that largely eliminated the cooling problem, but nothing could ease the vibration other than strengthening the affected structures, and this added more weight. The aircraft was briefly grounded in late 1946 for modifications, and then flew an additional 160 hours of tests by Air Force pilots. It then returned to Convair where it flew 117 more hours over 53 test flights using company test pilots.[10]

The XB-36 was designed with a single wheel and tire on each main landing gear, which was perhaps its biggest operational limitation. This resulted in the largest aircraft tire ever made (110-inch diameter x 36 inches wide) using enough rubber for 60 car tires.[11] Each tire weighed 1,475 pounds and each wheel was 850 pounds. The dual multiple disk brakes weighed 735 pounds. Total weight for each main landing gear including struts and ancillary equipment was 8,550 pounds. Only three airfields (Fort Worth, Eglin,

The XB-36 (42-13750) just after take-off featuring the huge single-wheeled landing gear, which limited the aircraft to only three existing airfields. (U.S. Air Force)

and Fairfield-Suisun) had the required 22-inch-thick runways that could support the 156-psi landing load. This issue was finally resolved when Major General Edward Powers,[12] Assistant Chief of Staff for Material, Maintenance, and Distribution, recommended that a new landing gear be developed. This new gear design resulted in a 4-wheel truck, which reduced weight by 1,500 pounds. It also eliminated the need for specially built runways, allowing the B-36 to use all B-29 airfields. However, this new landing gear was never installed on the first XB-36.

YB-36. The second XB-36 aircraft was subsequently redesignated YB-36. On 7 April 1945, the AAF designated the YB-36 (44-92005) as the production prototype due to the significant design issues on the first XB-36. The aircraft featured the bubble canopy that became the standard for B-36 production aircraft. Although originally designed with the single-wheel landing gear used on the XB-36, it was refitted with the four-wheel landing gear that was later used for production aircraft. This landing gear design significantly

The YB-36, and all subsequent aircraft, was fitted with a multi-wheel landing gear system which significantly increased the number of airfields available. (University of North Texas Libraries)

increased the number of airfields that could now accommodate the B-36. It also incorporated many of the design changes requested from the XB-36 design, such as reversible pitch propellers, nose gun turret provisions, moving No. 3 and 4 bomb bays adjacent to each other (instead of separated by the aft turret bays), and redesigned forward crew compartment. Two lower gun turrets were deleted to make room for the new AN/APG-3 tail radar, originally designed for the B-29, that replaced the AN/APG-7 Eagle.[13] The aircraft first flew on 4 December 1947 but was not delivered to SAC until October 1949 after spending nearly two years in testing. A year later it was returned to Convair for RB-36E conversion.

YB-36A. YB-36A (44-92004) first flew on 28 August 1947, which was six months before the YB-36. It made a short initial flight over Fort Worth and was formally delivered to the AAF on 30 August 1947. It was originally designated YB-36A[14] and was redesignated B-36A before delivery. It was fitted with just enough equipment for a flight to Wright Field where it was structurally tested to destruction based on the mid-1946 recommendation from Lieutenant General Nathan Twining. He said experience

The first B-36A was tested to destruction at Wright Field, OH after logging just 7 hours and 36 minutes flying time. (U.S. Air Force)

showed that the AAF would have been unable to use bombers effectively without this information. For example, the B-17, designed for 37,000 pounds gross weight, operated at 64,000 pounds. This could not have been done without accurate knowledge gained by static testing to destruction.[15] The aircraft logged just 7 hours and 36 minutes in its two flights before its destruction.

Bombers

B-36A. This was the first production B-36 model, but it was never considered an operational bomber. All B-36A aircraft were unarmed and used for crew training only. They were also limited to 72,000-pound bombloads and 310,380 pounds take-off weight due to landing gear restrictions. It incorporated the improvements made from YB-36 testing including the four-wheel landing gear, reversible pitch propellers, nose turret, and redesigned forward crew compartment. Although the AAF had approved replacement of the 3,000 horsepower R-4360-25 engine used on the XB-36 and YB-36 with the new 3,500 horsepower water-injection R-4360-41 engine it was not installed on the B-36A due to availability issues. The B-36A maximum speed was 345 mph at 31,600 feet with a 10,000-pound bombload.

B-36A (44-92006) parked on the ramp in Fort Worth on 5 January 1948. (University of North Texas Libraries)

A total of 22 aircraft were built. SAC took delivery of 20 B-36A aircraft and based them at Carswell AFB, TX. One of the 22 aircraft, B-36A (44-92004), originally designated YB-36A, was delivered on 30 August 1947 and flown directly to Wright Field, OH[16] for structural testing to destruction. Another, B-36A (44-92005), originally designated YB-36A, was delivered on 14 July 1948, and went to Wright Field where it was converted to EB-36A[17] and used for special testing. The next 11 aircraft were also originally designated YB-36A[18] but were subsequently redesignated B-36A on the production line before delivery.

Two aircraft were delivered in place to the Convair plant in Fort Worth and assigned to various activities before arriving at Carswell. B-36A (44-92006) was delivered on 3 June

1948 and initially went to Wright Field for testing on 7 June 1948, before being assigned to the 7th BG on 3 September 1948. The next B-36A (44-92007) was delivered at the Convair factory in Fort Worth on 18 June 1948, and then flew to Air Proving Grounds, Eglin AFB, FL on 9 July 1948 for testing before being assigned to the 7th BG on 23 July 1948. The first B-36A assigned directly to Carswell (44-92015), named "City of Fort Worth", was delivered to the 7th BG on 28 June 1948.[19] The 21 B-36A aircraft, including the EB-36A (44-92005) which was converted back to B-36A, and the YB-36 were later converted to 22 RB-36E models.

B-36B. This was the first operational B-36 model. A total of 73 aircraft were ordered but only 62 were delivered as B-36B. The others were converted at the factory (four delivered as B-36D and seven as RB-36D). The YB-36B (44-92026) first flew on 8 July 1948 and performed far better than expected. The B-36B was a clear improvement over the B-36A featuring the new R-4360-41 turboprop engines with water injection system,

Front view of B-36B (44-92056) parked on the tarmac at Convair Fort Worth. (University of North Texas Libraries)

which produced 500 horsepower more than the R-4360-25 engines used on the B-36A. Improved landing gear increased the gross weight to 328,000 pounds on tail numbers 44-92068 to 44-92070, 44-92076, and 44-92082 to 44-92087. All other aircraft were

B-36B (44-92033) in flight. (University of North Texas Libraries)

restricted to 278,000 pounds. The B-36B also had the "Grand Slam" modifications needed for carrying two Mk-17 hydrogen bombs, which weighed 42,000 pounds and had been created in such secrecy that Convair didn't have the dimensions in time for the A models.[20] The aircraft had a maximum bombload capability of 86,000 pounds. including two of the 43,000-pound T-12 Cloudmaker bombs. A total of 18 aircraft could carry two remote controlled VB-13 "Tarzon" bombs.[21] The aircraft also featured an improved AN/APQ-24 bombing navigation system.

B-36D. In January 1949, Convair submitted a proposal to mount two jet engines on pods under each wing. Tests had shown that bombers flying at high altitudes were difficult to locate and shoot down. Also, "putting on a burst of speed over the target" further increased survival. Convair reasoned that the jets could be used to supplement the B-

36 turboprop engines to achieve this burst of speed and improve overall performance when needed.[22] The YB-36D (44-92057) first flew on 26 March 1949. This aircraft was a converted B-36B with four Allison J35-A-19 jets installed because the planned J47-GE-19 engines were not yet available. These J35 engines failed to provide the thrust needed so Convair upgraded the aircraft to four J47 engines. A second B-36B (44-92046) was also equipped with J47 engines and used for high altitude testing. Neither of these aircraft incorporated the other changes planned for the B-36D, such as "snap-action" bomb bay doors, and they were among the last to be converted to the full B-36D configuration.[23]

The first B-36B (44-92057) aircraft converted to B-36D was originally designated YB-36D during initial testing. (University of North Texas Libraries)

A third B-36B (44-92090)[24] was fitted with J47 engines and used for an accelerated service test program. It flew more than 500 hours in the first 73 days of the program. The aircraft underwent a thorough inspection after every 120 hours of operation and flew tactical missions to evaluate how the aircraft behaved under extended operations at high altitudes and long ranges.[25] These flights proved the validity of this approach, in which the jets were only employed for takeoff, climbing to extreme altitudes, and dashing across hostile territory. The Air Force accepted it and authorized the conversion of B-36B aircraft to the new B-36D configuration. LeMay added his personal pledge: "I believe we can get the B-36 over a target and not have the enemy know it is there until the bombs hit." Even George Kenney came out of exile from his post as commander of the officer training center, Air University, to praise the airplane. "The B-36 went higher, faster, and farther than anybody thought it would," he said, "and the pilots liked it. It was a lucky freak." However, Kenney guessed that both the Navy Banshee and the Royal Air Force Vampire could intercept the B-36 in daylight; he recommended that it be used only on night raids.[26]

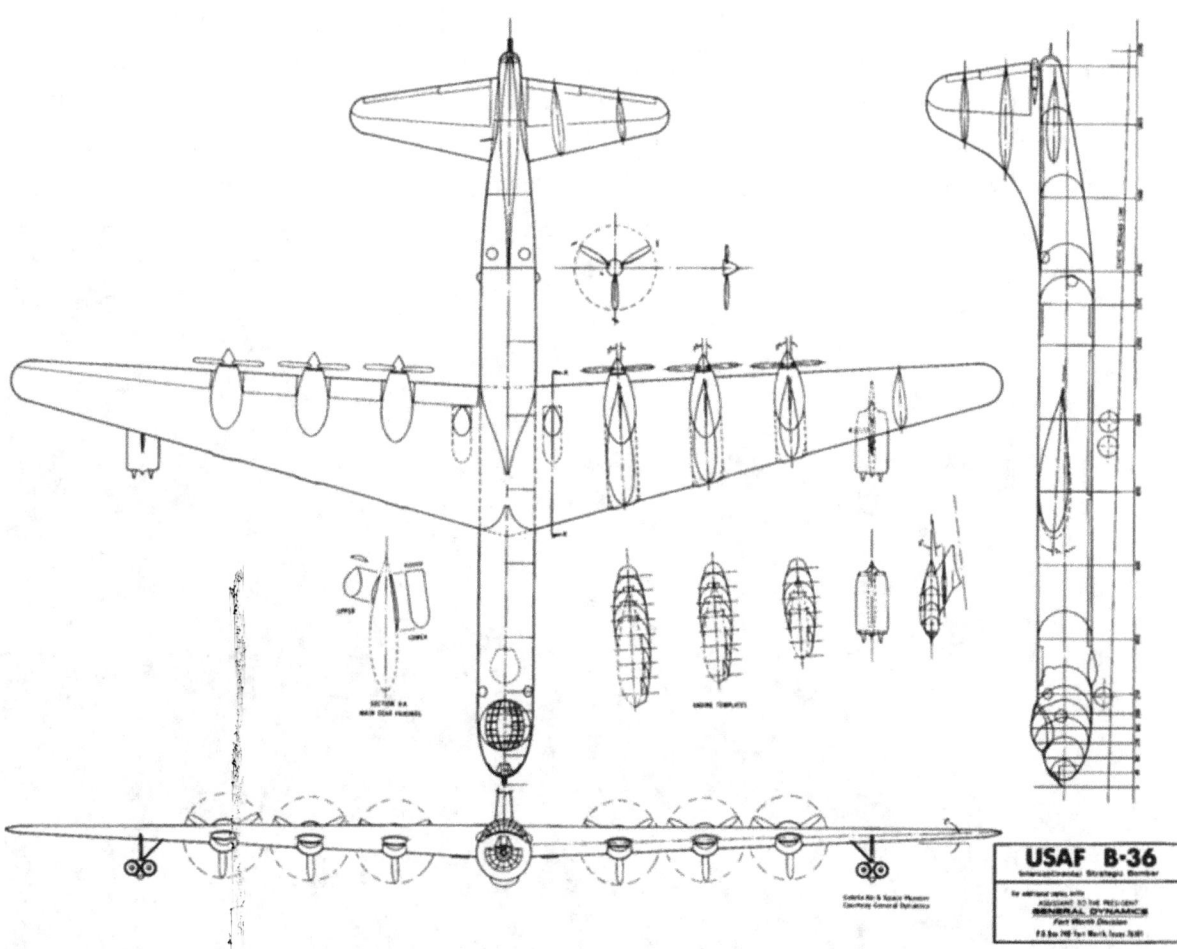

The B-36D aircraft featured many improvements over previous models including snap-action bomb bay doors and four J47 jet engines to improve the aircraft's speed over target. (U.S. Air Force)

The B-36D also featured several other improvements over the B-36B. The K-3A Bombing - Navigation System (BNS) replaced the AN/APG-24 radar and allowed one crew member to act as both the radar operator and bombardier. However, due to the unavailability of the K-3A, most aircraft were initially fitted with the K-1, which suffered from random vacuum tube failures and caused 25 percent of aircraft aborts. The AN/APG-32 radar replaced the AN/APG-3 radar to control the tail turret. Landing gear improvements increased the take-off and landing weights to 370,000 and 357,000 pounds, respectively. Finally, new "snap-action" bomb bay doors replaced the slow-moving hydraulically actuated doors used on the B-36A and B-36B allowing the doors to open and close in 2 seconds.[27]

The B-models were barely delivered when they were returned to Convair for B-36D conversion. The Air Force also accepted 22 B-36D (49-2647 to 49-2668) models as new-build production aircraft, and four B-36B (44-92095 to 44-92098) models converted during production and delivered as B-36D. The first delivery of a B-36B (44-92098) converted during production occurred on 18 August 1950. The first flight of a new-build production B-36D (49-2647) occurred on 11 July 1949 and it was delivered on 28 August 1950 after a year of testing.[28] The first delivery of a new-build production B-36D

Convair San Diego facility with B-36B to B-36D conversions in-work. (Consolidated Vultee Aircraft Corporation)

(49-2653) occurred on 22 August 1950. The Fort Worth plant was consumed with production of new-build aircraft so, after the first four conversions (44-92026, 44-92034, 44-92053, and 44-92054), the conversion effort was shifted to the Convair plant in San Diego, CA. The first conversion in Fort Worth (44-92054) was completed on 5 October 1950 and the first conversion in San Diego was completed on 5 December 1950.

B-36F. This aircraft was essentially a B-36D equipped with more powerful R-4360-53 engines, which produced 3,800 horsepower (300 more than the R-4360-41 engines used on the B-36D). The new engines included 25 improvements including direct fuel injection and new ignition systems for improved reliability at higher altitudes. Top speed increased to 417 mph and the service ceiling increased to 44,000 feet. There were no other noticeable differences from the B-36D other than minor changes in the cabin equipment arrangements. However, there were major changes in the defensive systems including the replacement of mechanical computers with new electronic units. Later production aircraft also featured two chaff dispensing systems to confuse enemy radar. The initial order consisted of 17 B-36F and 19 RB-36F aircraft as an add-on to the existing B-36D contract. The order was modified the following year to add 19 more B-36F and five RB-36F production aircraft.

The YB-36F (49-2669) first flew on 18 November 1950 and was delivered on 31 March 1951. However, three B-36F and three RB-36F aircraft (49-2670, 49-2671, 49-2672, 49-2703, 49-2704, and 49,2705) participated in a six-month accelerated test program at the Convair facility in Fort Worth before delivery. The program collected system information, operational and test data, and provided optimized training. Consequently, the first aircraft to enter service with SAC (49-2671) did not arrive until 18 August 1951. The final aircraft was accepted in January 1952.

B-36H. This was an improved version of the B-36F and retained the R-4360-53 engines. It featured a rearranged forward crew compartment including a second FE station. The radar was relocated to a pressurized compartment, and the AN/APG-41A defensive fire control system including twin tail radomes was installed (replacing the AN/APG-32 used on the B-36D and B-36F). The official first flight of the YB-36H (50-1083) occurred in November 1951.[29] B-36H deliveries began in place at the Convair Fort Worth plant in December 1951. However, assignments to SAC units were delayed due to a pressure bulkhead failure on an RB-36F, while flying at 33,000 feet, that restricted all B-36 aircraft to altitudes below 25,000 feet. SAC refused to accept deliveries of new aircraft until the entire fleet could be inspected and defective bulkheads replaced. Production B-36H aircraft finally entered service with SAC in beginning in May 1952.

B-36J. This was the last B-36 production model and was an improvement on the B-36H. It included two additional wing tanks to allow 2,770 gallons of additional fuel giving a total capacity of 36,396 gallons. The additional fuel load increased the combat radius

The flight crew stands in front of the last B-36 produced, a B-36J (52-2827), and delivered to the Air Force 16 August 1954. (University of North Texas Libraries)

by approximately 460 miles. It also included stronger landing gear that increased gross weight to 410,000 pounds. All B-36J aircraft were converted to B-36J-III Featherweight versions with the last 14 aircraft being completed on the production line before delivery. This reduced the crew to 13 and reduced weight by removing all defensive systems, except the tail cannon and its radar equipment, all crew comfort equipment, and other non-essential equipment. It increased the service ceiling to 47,000 feet and allowed some flights to reach well over 50,000 feet. The YB-36J (52-2210) first flew in July 1953. The aircraft entered service with SAC in September 1953 and was the most reliable model produced, since it benefitted from upgrades introduced in previous models.

Strategic Reconnaissance

RB-36D. The RB-36D was a standard B-36D fitted with reconfigurable sets of up to 14 cameras consisting of K-17C, K-22A, K-37, K-38, K-40, and T-11 for all-purpose strategic reconnaissance, day and night mapping and charting, and bomb damage assessment. It carried additional crew in a pressurized camera compartment installed in the forward bomb bay to operate and maintain the cameras. The compartment contained cameras and a small darkroom that allowed technicians to load and unload the film

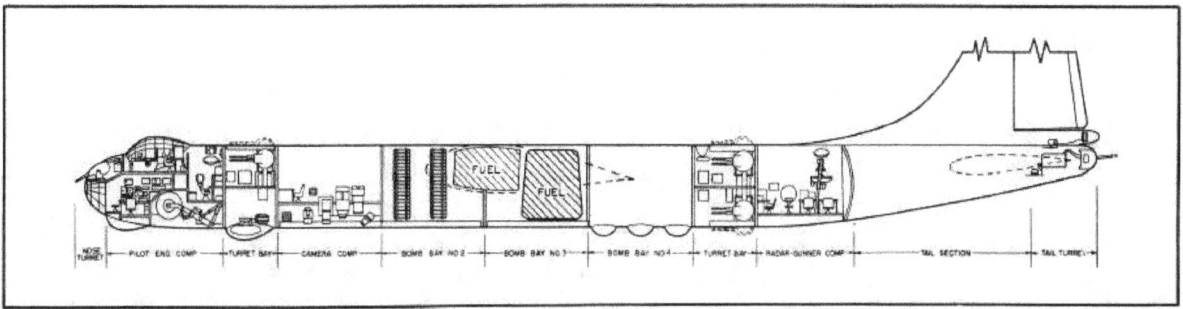

The RB-36D was a standard B-36D with a camera compartment installed in the No. 1 bomb bay including a dark room and up to 14 cameras. (U.S. Air Force)

during flight. The second bomb bay contained up to 80 T-86 photoflash bombs. The third bomb bay carried an extra 3,000-gallon droppable fuel tank which increased endurance to up to 50 hours. The fourth bomb bay carried Electronic Countermeasures (ECM) equipment. The aircraft was eventually modified to carry nuclear weapons in Bomb Bay No. 4.

A total of 24 aircraft were delivered including the first seven, which were ordered as B-36B (44-92088 to 44-92094) but converted during production and delivered as RB-36D

The RB-36D was fitted with reconfigurable sets of cameras for all-purpose strategic reconnaissance, day and night mapping and charting, and bomb damage assessment. (U.S. Air Force)

models. The aircraft first flew on 18 December 1949, and the Air Force began accepting aircraft six months later in June 1950 – two months before the first basic B-36D bomber. The final aircraft was accepted in May 1951. All 24 aircraft were assigned to the 28th SRW at Rapid City, SD. However, due to materiel shortages the RB-36D fleet did not become fully operational until June 1951. When developed, it was the only American aircraft having enough range to fly over the Eurasian land mass from bases in the United States, and size enough to carry the bulky high-resolution cameras of the day.[30]

RB-36D with its nose hoisted for a move on the assembly line in December 1950. (University of North Texas Libraries)

RB-36E. The RB-36E was developed as an immediate response to General LeMay's need for a strategic reconnaissance platform. A total of 22 aircraft were converted to RB-36E including 21 B-36A and one YB-36. Conversions began in early 1950 and the first aircraft (44-92024) was delivered to the 28th SRW at Rapid City, SD on 31 July 1950. The final delivery was in July 1951. Twelve aircraft were initially assigned to the 28th SRW and ten to the 5th SRW at Travis AFB, CA. All 22 aircraft were eventually transferred to the 72nd SRW at Ramey AFB, Puerto Rico.

The conversion included an upgrade of the engines from the R-4360-25 to the more powerful R-4360-41 engines used on the B-36B. They were also fitted with four J47 jet engines as used on the RB-36D. They were equipped with reconfigurable sets of 14 cameras, like the RB-36D, consisting of K-17C, K-22A, K-37, K-38, K-40, and T-11 for all-purpose strategic reconnaissance, day and night mapping and charting, and bomb damage assessment. Bomb bay configuration and crew complement were the same as the RB-36D.

RB-36D (49-2696) parked on the tarmac. (University of North Texas Library)

RB-36F. This aircraft was essentially an RB-36D equipped with more powerful R-4360-53 engines, which produced 3,800 horsepower (300 hp more than the R-4360-41 engines used on the RB-36D).[31] A total of 24 aircraft were produced. The first aircraft delivered (49-2707) was assigned to AMC on 8 May 1951 for testing of the Fighter and Conveyer (FICON) concept. It first flew in the FICON configuration in January 1952 and was later redesignated as JRB-36F. The first aircraft assigned to SAC (49-2703) was delivered to the 5th SRW at Travis AFB, CA on 28 May 1951. All aircraft were initially assigned to the 5th SRW, and later transferred to the 99th SRW at Fairchild AFB, WA and finally to the 72nd SRW at Ramey AFB, Puerto Rico.

RB-36H. This aircraft had similar performance to the RB-36F and incorporated most of the design improvements of the standard B-36H.[32] It was the most numerous of the reconnaissance variants with 73 aircraft produced. The first flight occurred in January 1952.[33] The first aircraft delivered to SAC (50-1104) was assigned to the 5th SRW at Travis AFB, CA on 7 February 1952. Although aircraft deliveries did not begin until 1952, the original contract was signed in 1950 resulting in the aircraft carrying 1950 series tail numbers.

RB-36H aircraft undergoing maintenance on the flightline at Carswell AFB, TX. (U.S. Air Force)

Special Mission
DB-36H. On 7 July 1952, the Air Force issued a contract to Convair to develop a prototype Guided Air Missile (GAM) Director aircraft able to carry and direct the GAM-63

RASCAL Nuclear Stand-Off Missile

The GAM-63 RASCAL was the first nuclear standoff missile carried aboard a bomber. It was powered by a 4,000-pound liquid-fueled rocket engine made up of three vertical in-line thrust chambers. The missile was 31 feet long with a body diameter of four feet and could carry a 3,000-pound nuclear warhead up to 110 miles at a maximum speed of Mach 2.95.

The RASCAL was fitted with a retractable antenna to provide a datalink to the bomber. The B-36H modification was designed to allow the aircraft to convert back to the basic bomber configuration within 12 hours. However, Convair demonstrated that this could be completed in 3 hours and 12 minutes. The RASCAL was loaded into the aft bomb bays and the guidance package was loaded into the No. 1 bomb bay. The missile was carried semi-submerged in the combined No.3 and No.4 bomb bays and hung down about 18 inches. The inside of the bomb bay was protected by a fiberglass cover after the missile was released.

RASCAL was the first nuclear stand-off missile carried by a bomber. (U.S. Air Force)

Modified controls at the bombadier's station allowed him to launch the missile from the bomb bay as the aircraft neared the target. Once it was released it followed a preprogrammed flight path but the bombardier, acting as the missile operator, could make minor adjustments through the datalink.

RAdar SCAnning Link (RASCAL) missile. The production order for 11 more DB-36H aircraft was exercised on 26 May 1953. The prototype designated YDB-36H (51-5710) made its first flight on 3 July 1953. However, delays in the development of the RASCAL and its guidance system quickly delayed the program by one year. The final schedule called for two deliveries in late 1955 and eight more before 15 November 1956.

Ultimately, only three aircraft were built. In addition to the prototype, B-36H (51-5706) was released after the termination of the Tanker-Bomber (Tanbo XIV) program (See B-36H Tanker) and converted to DB-36H GAM Director. It was redesignated EDB-36H to designate special test status and then redesignated JDB-36H in 1955. A third B-36H aircraft (50-1085) was converted to the GAM Director configuration and designated DB-36H. It was later redesignated EDB-36H.

By July 1955, the Air Force had launched a RASCAL from a DB-47E and decided to stop production of the DB-36H. On 9 September 1958, the RASCAL program was considered obsolete and terminated due to other technological advances.

GRB-36D. A total of ten RB-36D aircraft (44-92090, 44-92092, 44-92094, 49-2687, 49-2692, 49-2694, 49-2695, 49-2696, 49-2701, and 49-2702) were converted to the GRB-36D configuration. These aircraft were modified to carry a GRF-84F Thunderstreak on a ventral trapeze as part of the FICON program (See JRB-36F). The GRF-84F was carried partially withdrawn into the bomb bay and launched and retrieved by means of a retractable boom.[34]

Experiments

B-36C. The B-36 was criticized throughout its service life for its slow speed, and Convair constantly looked for ways to improve it. In March 1947, they proposed a new configuration using R-4360-51 Variable Discharge Turbine (VDT) engines. This configuration would require a significant redesign of the wing structure to allow the engines to face forward and use forward facing (Tractor vs. Pusher) propellers. The proposed design required the engines to remain in the same location on the wing trailing edge and a ten-foot shaft would drive the propellers on the leading edge. Convair expected to increase the aircraft's top speed to 410 mph with a 45,000-foot service ceiling and 10,000-mile range. They proposed that 34 of the 100 aircraft already on order be built in this new configuration, designated B-36C. They also proposed financing the prototype structural design changes by removing three production aircraft from the 100-aircraft order.[35] The new Aircraft and Weapons Board met on 19 August 1947 and determined that the B-36 would remain in service only as an interim atomic bomber and eventually be replaced by the B-52. They decided there was no operational need for the B-36C, and it was cancelled on 22 August 1947.

But Convair persisted and submitted a new proposal for the B-36C on 4 September 1947. This time they proposed removing five aircraft from the 100-aircraft order and said they would deliver the YB-36C prototype in November 1948. There was stiff opposition within the Air Force to the B-36C, and any other production version of the B-36 beyond the 100 already ordered. There were concerns over the availability of the VDT engines that were also being used for the B-50C and other aircraft, potential cost for government furnished equipment not covered by the reduction in aircraft production, and SAC's opposition to the B-36 in general. Despite these concerns, the Aircraft and Weapons Board met again in November 1947 and approved the Convair proposal for 34 aircraft on 5 December 1947. However, due to cooling problems caused by how the VDT engine was installed in the B-36C, the aircraft never achieved the speeds promised and the order was cancelled in the spring of 1948. The 34 aircraft were built as B-36B models instead.

B-36H Tanker. In late 1951, SAC began looking at the concept of a B-36 convertible Tanker-Bomber (Tanbo) capable of refueling the new B-47 bombers and fighter aircraft. The Air Force issued a contract to Convair on 15 January 1952 to install a probe and drogue refueling system including the Mark XIV refueling reel. One B-36H (51-5706) was modified and was known as Tanbo XIV. The refueling reel and hose, transfer pumps, and hydraulically operated boom that extended into the airstream and dispensed the hose were mounted in the No. 4 bomb bay. Fuel bladders were installed in the remaining bomb bays. The hose reel could be removed from the No.4 bomb bay in about 12 hours allowing an atomic bomb to be loaded, but the fuel bladders were permanently installed preventing bomb loading in the remaining bomb bays. The system allowed 16,000 gallons of fuel offload, at 600 gallons per minute, allowing two or three B-47 bombers to top off on their way to target. Testing began in March 1953 but soon ran into problems with the reel mechanism, hose, and boom. Testing was suspended on 27 May 1953. The problems were not resolved before the KC-135 tanker became reality. The aircraft was demodified and transferred to the DB-36H GAM Director aircraft program on 6 July 1954. The Tanbo XIV program was subsequently cancelled on 21 Jul 1954.[36]

ERB-36D. The first RB-36D (44-92088) spent its entire service life as a test aircraft. The standard RB-36D carried up to 14 cameras in varying configurations of K-17C, K-22A, K-37, K-38, K-40, and T-11 cameras. A special 240-inch focal length K-42 camera (known as the Boston Camera after the university where it was designed) was tested on 44-92088, and the aircraft was redesignated ERB-36D in 1954. The long focal length was achieved by using a two-mirror reflection system. The camera could resolve a golf ball at an altitude and side range of 45,000 feet. That is a slant range of over 63,600 feet.[37] The camera was tested

The Boston Camera with its 240-inch focal length could photograph a golf ball from 45,000 feet - and this was in the 1950s! (McChizzle)

for about a year and the Air Force determined that the $200,000 required was too costly to return the aircraft to production standard after the camera was removed. The aircraft was scraped instead at Kelley AFB, TX in late 1955.[38]

EB-36H. Four B-36H aircraft (51-5726, 51-5731, 52-1357, and 52-1358) were redesignated EB-36H and were assigned to the 4925th Test Group (Atomic) at Kirtland AFB, NM to support atomic test operations. These aircraft were redesignated JB-36H in 1955 when the Air Force started using the "E" designation for electronic systems equipped aircraft. These aircraft participated heavily in atomic testing operations. They were used to measure the effects of nuclear explosions on airborne aircraft as well as to measure radiation released from nuclear bombs.

JB-36D. One B-36D (44-92054) was redesignated JB-36D and used for special test and development.

JB-36F. One B-36F (49-2677) was redesignated JB-36F and used for special test and development. This aircraft was used to carry a B-58 static test aircraft attached to its belly from Fort Worth to Wright-Patterson AFB, OH on 12 March 1957. The inboard propellers had to be removed to avoid hitting the B-58.

RB-36D (44-92088) under construction in November 1949. It was later modified to carry the "Boston Camera" and redesignated as ERB-36D. (University of North Texas Library)

JRB-36F. Beginning in 1951 the Air Force was exploring ways to protect RB-36 aircraft, which were becoming more vulnerable to enemy fighter attacks and missiles. They envisioned carrying a parasite fighter on the RB-36 that could be released about 800-1,000 miles from the target to defend the bomber or perform reconnaissance missions on its own. This would not only provide the bomber with its own fighter protection but would make it possible for the bomber to carry the fighter long distances to a combat zone. The concept proved successful, and the Air Force awarded a contract in the fall of 1953. This concept was known as the FICON program. A single RB-36F (49-2707) was modified as the FICON prototype and made its first flight in January 1952. It was later redesignated as JRB-36F on 30 October 1956. Ten additional FICON aircraft were converted from RB-36Ds for use as fighter conveyors and redesignated as GRB-36D. However, subsequent development of mid-air refueling proved so successful that experiments with parasite fighters were discontinued.[39]

YRF-84F FICON below the JRB-36F. (National Museum of the U.S. Air Force)

JRB-36H. Two RB-36H aircraft (51-5748 and 51-5750) were converted to JRB-36H and assigned to the 4925th Test Group (Atomic) at Kirtland AFB, NM to support high-altitude atomic tests. They were modified with an upward looking camera to photograph mushroom clouds during Operation HARDTRACK I at Eniwetok and Bikini Islands in 1958.

NB-36H. One B-36H (51-5712) damaged beyond repair during the 1 September 1952 tornado at Carswell AFB, TX was fitted with the world's first airborne nuclear reactor installation for trials. It was originally designated XB-36H and subsequently NB-36H. It had a redesigned cockpit to shelter the crew from the reactor and a raised nose. The plane was conventionally powered but employed the "hot" reactor to test radiation effects on equipment and crew shielding. The first flight was made on 20 July 1955.[40] This aircraft was intended to evolve into the Convair X-6.

The NB-36H was fitted with an operational nuclear reactor to test the effects of radiation. (U.S. Air Force)

YB-60. This was a project for an eight-engine jet-powered swept wing variant. Two aircraft (49-2676, 49-2684) were pulled from the B-36F production order and delivered as YB-36G jet bomber prototypes. Due to its significant differences from a standard B-36 the designation was later changed to YB-60. However, it still retained approximately 72 percent of parts common with the B-36. The fuselages were almost identical, but the YB-60 had a more pointed nose with an instrument probe. The wings were swept-back, and eight engines were hung on four engine pods like the B-52. The tail and horizontal stabilizers were also more streamlined than the B-36. It became a competitor to the B-52 but, although it was a larger aircraft and could carry a heavier bombload, it was too slow compared to the B-52. The program was cancelled in January 1953. Although the first aircraft (49-2676) flew several test flights, the second aircraft (49-2684) never flew. Both aircraft were scrapped in July 1954.

The YB-60 was an 8-engine all-jet bomber that was 72 percent compatible with the B-36 bomber. (U.S. Air Force)

XC-99. This was a cargo/transport version of the B-36. One was built (43-52436) and first flew on 23 November 1947. It was the world's largest airplane at the time, built on

XC-99 and B-36 in-flight. (National Museum of the U.S. Air Force)

the B-36 baseline of the largest piston aircraft ever built. The aircraft flew in active service and accumulated 7,400 flying hours and moved more than 60 million pounds of cargo. It made its last flight on 19 March 1957.

Featherweights

In 1954, the Air Force began the Featherweight modification program designed to increase the altitude and range of the B-36. The goal was to eliminate the need for the pre-strike staging at northern bases, mostly outside of the United States, required for the B-36 to reach targets in the Soviet Union. The Featherweight modifications required the removal of unnecessary equipment that significantly reduced the aircraft weight. As reflected below, these changes resulted in an average decrease across the fleet of 3% combat weight, and an average increase in combat radius, max speed, and service ceiling of 11%, 2%, and 2% respectively.

	Combat Weight (pounds)		Combat Radius (miles)		Max Speed (mph)		Service Ceiling (feet)	
Model	Basic	FW III	Basic	FW III	Basic	FW III	Basic	FW III
B-36D	248,410	244,400	3,529	3,754	407	418	44,300	45,600
RB-36D	258,675	251,900	3,494	3,714	401	415	43,400	44,200
RB-36E	258,200	251,900	3,520	3,714	401	415	43,400	44,200
B-36F	254,300	248,400	3,232	3,673	418	421	44,000	44,500
RB-36F	262,800	256,400	3,098	3,552	409	417	43,100	43,750
B-36H	253,900	241,300	3,113	3,655	416	423	44,000	45,700
RB-36H	263,300	254,600	2,936	3,622	408	417	42,700	43,900
B-36J	266,100	262,500	3,403	3,990	411	418	43,000	43,600

Effects of Featherweight III changes across the fleet.

The Featherweight modifications resulted in the following configuration changes.[41]

Featherweight I. This involved the owning unit removing all retractable turrets, auxiliary bomb racks, and crew comfort items before a retaliatory strike. It was never implemented due to a potential several-day delay in missions. However, owning unit maintenance personnel completed general housekeeping by removing unnecessary brackets, spare parts, etc. during routine maintenance.

Featherweight II. This modification was conducted at the depot and involved removing extraneous equipment except guns, ECM equipment, auxiliary bomb racks, and crew comfort items. The gun turrets were slightly reconfigured to allow quick removal. External perturbances were removed, when possible, to reduce drag. These modifications reduced gross weight by 4,800 pounds. Flush covers were procured for all six blisters. If needed, the owning unit could quickly remove the guns and install the flush panels to further reduce weight by several thousand pounds.

Featherweight III. This modification was also completed at the depot. It involved removing all guns and their equipment, except the tail gun and its radar equipment. The forward and upper blisters were covered with flush panels. The aft lower blisters were kept open to allow observation of the engines, although for some aircraft a plexiglass flush panel was installed in place of the blister panel. Most of the crew comfort equipment was removed and the crew was typically reduced by two gunners. Most insulation was removed, and the aircraft was fitted for heated flying suits. The modifications reduced gross weight by 15,000 pounds.

SAC plans called for some of the aircraft to retain the Featherweight II configuration for missions where guns were required. Other aircraft received the Featherweight III configuration to maximize their range.

With its enormous wings and extra fuel, who knows how high and how far it could fly? B-36 crews speak of 45-hour missions, presumably with fuel cells instead of nukes in the rear bomb bays; at cruise speed, a "Featherweight" could travel almost 9,000 miles in that period. The official ceiling was 41,300 feet, but again, crews say that they routinely flew higher than 50,000 feet, and one man – John McCoy, quoted in Thundering Peacemaker – boasted of soaring to 58,000 feet. On missions over China, McCoy said, his RB-36 was chased by MiG fighters that could not climb anywhere near it. United States fighter pilots of that period also recall B-36s cruising comfortably well above their own maximum altitude. Not until the advent of the "century series" fighters – the F-100 and up – would the B-36 be challenged. Whether the RB-36 ever overflew Russia is anyone's guess, but it was the United States altitude and distance champ until the Lockheed U-2 came online toward the end of the decade.[42]

Organization and Basing

At the time of the Japanese surrender from WWII, the AAF had 218 operational groups. The War Department's postwar planning called for a "bedrock minimum" air force of 70 groups and a personnel strength of 400,000. In February 1946, General Arnold retired, and General Spaatz took over as Chief of Staff. He held to the objective of 70 air groups and was supported in that position by the Army Chief of Staff, General of the Army Dwight D. Eisenhower. However, the 70-group plan evaporated in August 1946 when President Truman ordered "economy commitments" to reduce the budget by $2.2 billion, with 75 percent of that to come from the armed forces. At the end of 1946, the Air Force had 55 groups, only two of which could be counted as combat ready. "We are not ready to fight a war if one came today—and we won't be for quite a long time," Spaatz said. The Air Force had not yet reached the bottom. It would sink to 48 groups and a personnel strength of 304,000 before the buildup for the Korean War began.[1]

It was a shock for the United States to learn that its wartime ally, the Soviet Union, had turned into an adversary. When former British Prime Minister Winston Churchill warned that an "iron curtain" had descended on Europe, many Americans refused to believe it. In the atomic age, the United States relied increasingly on the AAF, already moving toward its new status as a separate military service. In January 1946, the AAF was down to 30 percent of its wartime strength. By summer, the inventory of aircraft had been reduced by half, most of them cut up for scrap.[2]

When Spaatz established SAC on 21 March 1946, it inherited a force of approximately 1,300 bomber, fighter, reconnaissance, and support aircraft. The initial SAC force included 100,000 personnel with 22 major installations and 30 minor bases. With postworld demobilization in full swing this force was quickly reduced by the end of 1946 to a remaining force of 279 aircraft including 148 B-29, 85 P-51, 31 F-2, and 5 C-54. Personnel were reduced to 37,092 with only 18 bases remaining.

SAC's bomber force was organized into nine bomb groups, designated Very Heavy (VH), comprised of B-29 bombers. Six of these groups had 30 UE aircraft each and three had no aircraft assigned. When SAC began receiving B-36 aircraft in 1948 it realized there was a conflict with its previous designation of B-29 aircraft as VH compared to the comparatively massive B-36. Consequently, it redesignated its bomber force based on range rather than weight. Aircraft with a combat radius greater than 2,500 miles were designated Heavy (H), aircraft with an operating radius 1,000 – 2,500 miles were designated Medium (M), and less than 1,000 miles were designated Light (L). This new approach designated the B-29 as (M) and the B-36 as (H). Their associated bomb groups carried the same designations.

Determining the proper basing approach for bedding down the B-36 was a major focus for SAC. Although redesign of the single wheel main landing gear used on the XB-36 to

B-36 aircraft on the tarmac at Carswell AFB, TX. (American Aviation Historical Society)

a four-wheel truck allowed the B-36 to land at SAC's B-29 bomber bases, facilities were a key issue. The massive size of the B-36 compared to the B-29 bombers previously hosted at these bases meant suitable hangars were unavailable. Additionally, most B-29 bases were established in southern states during WWII due to the relatively mild weather that was essential for training aircrews. This became an issue during the Cold War, when SAC targeted locations in the Soviet Union, and the range from these southern bases exceeded the bomber's capability.

In 1948 SAC established its first two B-36 (H) bomb groups with 18 UE aircraft each. In March, SAC announced that the 7th BG at Carswell AFB, TX would be the first operational B-36 unit. Carswell was a logical choice since it was directly across the runway from Convair's Forth Worth assembly plant where all B-36 aircraft were built. Not only did it make delivery as simple as towing the aircraft across the runway, but it gave Convair access to the fleet for modifications and configuration changes and it gave the Air Force ready access to Convair training and engineering. The 11th BG at Carswell would be the second B-36 unit. SAC subsequently fielded its B-36 and RB-36 aircraft at eight bases across the United States and in Puerto Rico.

Beginning on 17 November 1947, the 7th Bomb Wing (BMW) was formed as a test unit for the newly developed "Wing-Base" plan, also known as the "Hobson Plan", in which the wing became the predominant operational unit on each base. The wing commander became the operational commander, and all combat and support groups reported to the wing. A base commander had responsibility for base housekeeping and administrative functions. The test was successful, and the 7th BMW was formally activated on 1 August 1948. Both the 7th and 11th BG operated as components of the 7th BMW until 1 February 1951 when the 11th BG was redesignated as the 11th BMW.

This designation of the 11th BMW at Carswell was part of the reorganization of SAC's combat command structure, based on a plan devised by General LeMay, using experience gained from Korean War operations. LeMay wanted an organization designed for peacetime that required minimum change to rapidly switch to war footing. Previously, despite having a base commander assigned, the wing commander was still occupied with too many base housekeeping and administrative tasks that distracted from his ability to focus on combat operations. Also, the wing commander relied heavily on the bomb group commanders assigned to the wing to conduct daily operations. With the reorganization, the Air Base Group commander assumed all base housekeeping and administrative duties. Although the combat groups continued to exist, they became subordinate to the wing and the wing commander served as the group commander. This gave full control and focus of combat operations to the wing commander.

In conjunction with this reorganization, SAC received authority from Headquarters Air Force to organize Air Division (AD) headquarters on double wing bases and to operate only one Air Base Group on these installations. The AD commander, typically a 1- or 2-star general officer, exercised direct control over the two wing commanders and the Air Base Group commander. For example, on 16 February 1951 the 19th AD was established at Carswell to oversee operation of the two wings assigned to the base.[3] SAC fully implemented this change across the command over the following year. On 16 June 1952, all SAC combat groups were eliminated, and the combat squadrons reported directly to the wing. All SAC wings and ADs reported directly to a Numbered Air Force (NAF) commander.

Unit	Components	Base	Aircraft	Notes
5th SRW (H) (1951-55) 5th BMW (H) (1955-58)	5th SRS/BS 31st SRS/BS 72nd SRS/BS	Travis AFB, CA	RB-36D (Jan 51 - Jun 53) RB-36E (Feb 51 - Jun 53) RB-36F (Aug 51 - May 54) RB-36H (Feb 52 - Sep 58)	Only one RB-36D assigned between Jan 51 and Jun 53. Performed strategic reconnaissance mission until 1955; shifted to bombardment training mission beginning in 1954.
6th BMW (H) (1952-57)	24th BS 39th BS 40th BS	Walker AFB, NM	B-36F (Aug 52 - Aug 57) B-36H (Oct 52 - May 54) B-36J (May 54 - Jun 57)	
7th BG (H) (1948-51) 7th BMW (H) (1951-58)	9th BS 436th BS 492nd BS	Carswell AFB, TX	B-36A (Jun 48 - May 49) B-36B (Nov 49 - Jun 51) B-36D (Aug 50 - Dec 53) B-36F (Aug 51 - Dec 52) B-36H (May 52 - Jun 58) B-36J (Nov 53 - Feb 58)	7th BG operated as component of 7th BMW until 1951.
9th BMW (H) (1949-50)	1st BS	Fairfield-Suison AFB, CA (Renamed Travis AFB in 1951)	B-36B (Nov 49 - May 50)	
11th BG (H) (1948-51) 11th BMW (H) (1951-52)	26th BS 42nd BS 98th BS	Carswell AFB, TX	B-36A (Jan 49 - Jul 49) B-36B (Jan 49 - Jun 51) B-36D (Apr 50 - May 54) B-36F (Aug 51 - Feb 53) B-36H (May 52 - Nov 57) B-36J (Nov 53 - Aug 57)	11th BG operated as component of 7th BMW until 1951.
28th BG (H) (1949-50) 28th SRW (H) (1950-55) 28th BMW (H) (1955-58)	5th SRS/BS 31st SRS/BS 72nd SRS/BS	Rapid City AFB, SD (Renamed Ellsworth AFB in 1953)	B-36B (Jul 49 - Oct 50) RB-36D (Apr 50 - May 53) RB-36E (Jul 50 - Jun 53) RB-36H (Mar 52 - Nov 58)	Performed strategic reconnaissance mission until 1955; shifted to bombardment training mission beginning in 1954.
42nd BMW (H) (1953-56)	69th BS 70th BS 75th BS	Limestone AFB, ME (Renamed Loring AFB in 1954)	B-36D (Feb 53 - Apr 56) B-36H (Feb 53 - Sep 56) B-36J (Sep 54 - May 56)	
72nd SRW (H) (1952-55) 72nd BMW (H) (1955-59)	60th SRS/BS 73rd SRS/BS 301st SRS/BS	Ramey AFB, Puerto Rico	RB-36D (Jan 53 - Sep 56) GRB-36D (Feb 55 - Mar 55)* RB-36E (Oct 52 - Feb 57) RB-36F (Jun 56 - Oct 58) RB-36H (Feb 57 - Nov 58)	Performed strategic reconnaissance mission until 1955; shifted to bombardment training mission beginning in 1954. *Temporarily assigned before assignment to 99th SRW.
92nd BMW (H) (1951-56)	325th BS 326th BS 327th BS	Fairchild AFB, WA	B-36D (Jul 51 - Mar 57) B-36J (Jun 54 - Feb 58)	
95th BMW (H) (1953-59)	334th BS 335th BS 336th BS	Biggs AFB, TX	B-36D (Jan 53 - Sep 56) B-36H (Feb 56 - Feb 58) B-36J (Feb 56 - Feb 59)	
99th SRW (H) (1951-55) 99th BMW (H) (1955-56)	346th SRS/BS 347th SRS/BS 348th SRS/BS	Fairchild AFB, WA	RB-36D (Dec 52 - Apr 55) GRB-36D (Feb 55 - Jul 56) RB-36F (Dec 52 - Oct 56)	Performed strategic reconnaissance mission until 1955; shifted to bombardment training mission beginning in 1954.

SAC fielded its B-36 and RB-36 aircraft at eight bases across the United States and in Puerto Rico.

SAC employed up to three NAFs during its history, the 2nd, 8th, and 15th. Initially, medium and heavy bomber units were assigned to 8th AF while 15th AF had mostly medium bombers and 2nd AF concentrated on reconnaissance. This arrangement resulted in an unbalanced organization with wings in the same geographical region being assigned to different NAFs. On 1 April 1950, SAC reorganized the NAFs on a geographical basis with 2nd AF in the eastern, 8th AF in the central, and 15th AF in the western United States. SAC further reorganized its NAFs on 13 June 1955 as SAC operations expanded in the Northeastern United States to support Cold War operations. The 2nd AF was now responsible for the southeast (including Texas), 8th AF was responsible for the northeast and central United States, while the 15th AF had the southwest and west.

Unlike current-day intercontinental bombers the B-36 was incapable of aerial refueling, making it the only true intercontinental bomber. Aircraft operating from Carswell and other "Southern Tier" bases had to forward deploy to Thule AB, Greenland, Goose Bay, Labrador, and Alaska to allow them to fly over the north pole and hit targets in the Soviet

B-36 aircraft on a snowy ramp at Limestone AFB, ME (later renamed Loring AFB). (American Aviation Historical Society)

Union. As the B-36 role as a nuclear bomb delivery platform and the growing Cold War with the Soviet Union became more apparent, the need for "Northern Tier" basing became clear. In 1947, SAC began construction of a new base in Limestone, ME built from the ground up to accommodate 60 B-36 bombers.[4] In 1950, SAC established the 28th SRW at Rapid City AFB, SD. In 1951, SAC established the 92nd BMW and 99th SRW at Fairchild AFB, WA and in 1953 the 42nd BMW at Limestone AFB, ME. SAC bombers could now reach targets in the Soviet Union from its United States bases. These bases also allowed additional forward operating locations for the southern-based wings.

SAC also continued to rely on forward operating bases in Alaska, Goose Bay, and Thule to shorten the distance to targets in the Soviet Union as shown below. However, these bases presented significant challenges to B-36 operations due to the extreme cold conditions. The B-36 did not respond well to cold weather and the aircraft had to be preheated before each flight. Furthermore, the facilities at these locations and all SAC northern tier bases were inadequate for the B-36 due to its size.

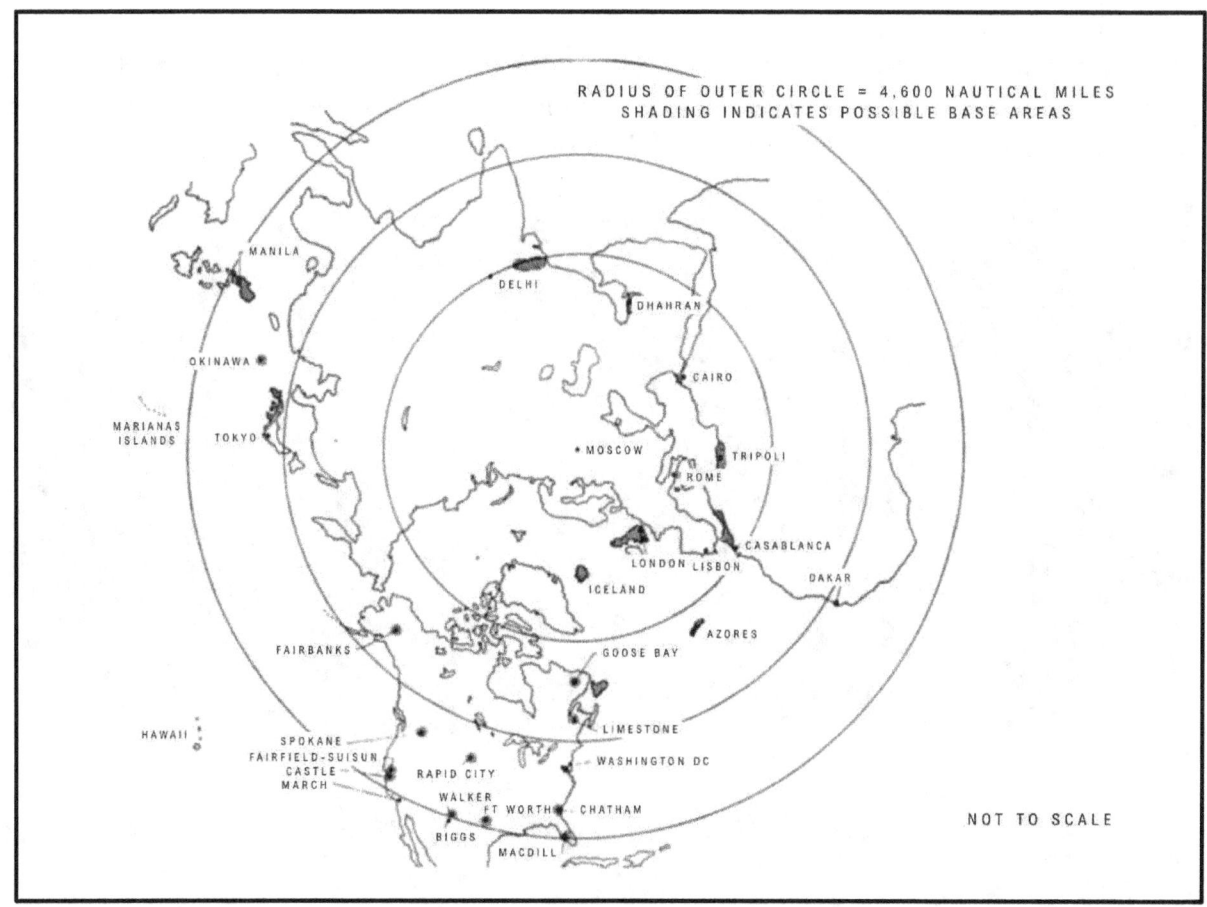

SAC B-36 basing plan allowed southern tier units to forward deploy to Alaska, Labrador, and Greenland to put them within reach of Soviet targets (History of AMC Support to Strategic Air Command 1946-1952).

SAC addressed this issue by equipping their northern tier bases with new "nose dock" hangars which allowed the aircraft, except the tail, to be indoors out of the harsh weather for maintenance. Both Limestone and Rapid City also got new "Arch Hangars" which fully enclosed two B-36 aircraft.[5] These hangars were built in 1947-48 and used a thin-shell concrete design in which huge concrete arches created a huge two and half acre open floorspace. SAC intended to build up to 10 of these hangars at Limestone and similar numbers at other northern tier bases. However, this plan was downscaled and eventually abandoned due to a shortage of concrete and other materials in the wake of the post-WWII building boom.

Nose dock hangars built at Loring AFB, ME and most other SAC bases allowed maintenance crews to work on B-36 Bombers with most of the aircraft sheltered from severe weather. (JC Garbinski)

B-36 aircraft undergoing maintenance in a nose dock hangar at Loring AFB, ME. (Loring Air Museum)

Instead, beginning in the early 1950s, SAC settled on a new Double-Cantilever (DC) hangar design capable of fully enclosing, except the tail, two B-36 aircraft in each bay. The DC hangar could be constructed with up to three bays capable of housing five B-36 aircraft total. Three-bay DC hangars were constructed to accommodate B-36 maintenance at Biggs, Carswell, Loring, and Walker and at several other bases to accommodate other SAC aircraft. Two-bay DC hangars were built at other SAC bases and supplemented with nose-dock hangars.

Loring AFB Arch Hangar capable of fully enclosing two B-36 bombers with the DC Hangar capable of holding five partially enclosed B-36 aircraft in the background. (JC Garbinski)

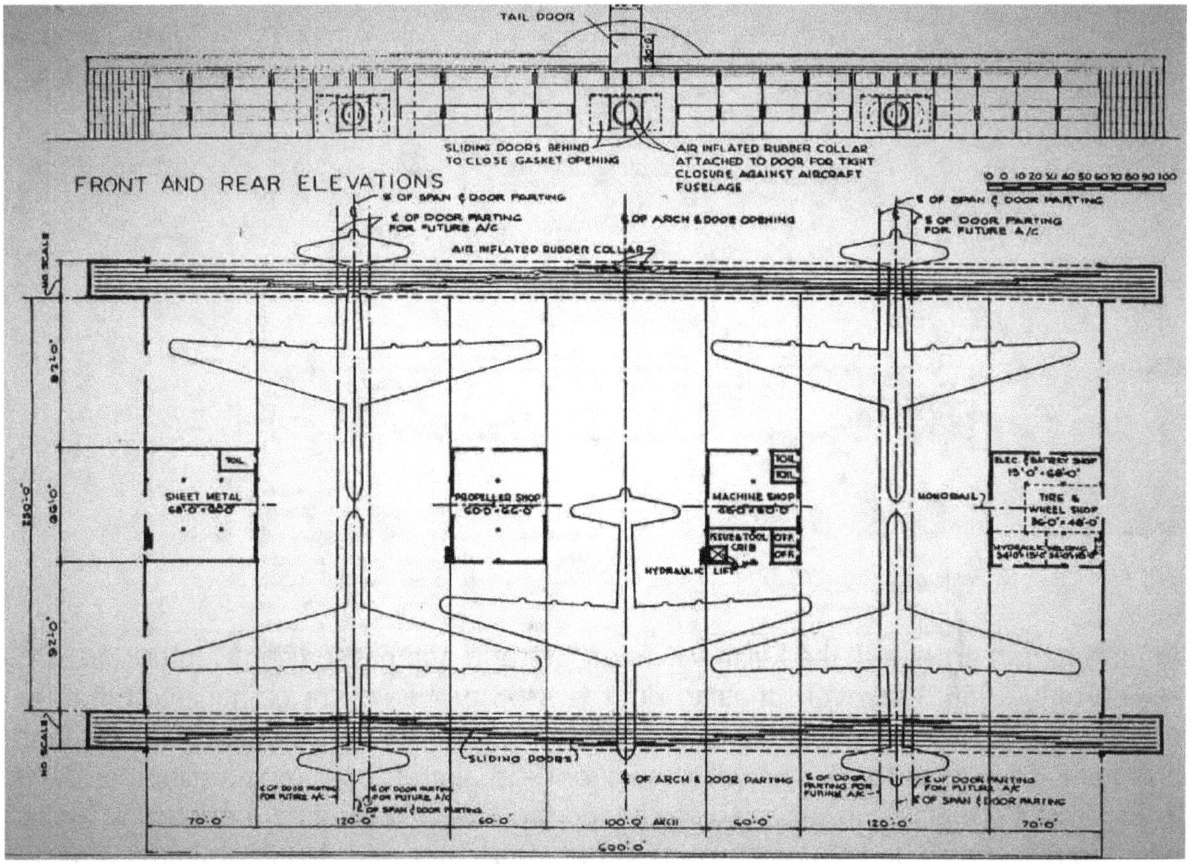

DC hangar could house up to five B-36 aircraft simultaneously. (U.S. Air Force)

SAC also began to develop forward operating locations in warmer climates such as the United Kingdom (UK), Morocco, and Guam. These locations put B-36 aircraft within range of the Soviet Union and were much more operationally suitable. B-36 aircraft flying from Guam were only 6,100 miles from Moscow and 2,200 miles from Vladivostok putting most targets in the Soviet Union in range.

In 1955, SAC redesignated its SRW units to BMW units. Under this organizational change the RB-36 aircraft were converted to bombers and retained some latent reconnaissance capability. The primary mission became strategic bombardment training. The reconnaissance mission was largely passed to newly formed RB-47 units and the still secret U-2. In 1956, SAC began phasing out its B-36 fleet. It had originally planned to start replacing them with B-52 bombers beginning in 1953, but defense spending cutbacks and technological issues stretched out B-52 procurement and caused the B-36 service life to be extended.

B-36 from the 7th BMW shown with the YB-52 at Carswell AFB, TX in 1955. (U.S. Air Force)

Aircraft began arriving at the Military Aircraft Storage and Disposition Center (MASDC) in February 1956. However, phasing out the B-36 proved more complicated than expected since SAC needed to maintain its global nuclear delivery capability. Initial shortages of B-52 aircraft forced the withdrawal of B-36 aircraft from reclamation contracts. B-36 aircraft remaining in service were supported with parts from out of service aircraft until the B-36 could be completely phased out. Scrapping the first 200 aircraft returned

$93.5M to the Air Force and approximately $88M more was returned due to engines parts that could be used on other Air Force aircraft.[6]

B-36 aircraft lined up at MASDC beginning in 1956 to await their demise. (U.S. Air Force)

The B-36 fleet was finally phased out by 1959 but remained first-line strategic bombers up to the final day, and they often flew from their last operational mission directly to MASDC. Although SAC eventually developed response times of minutes for fully loaded B-52 alert aircraft to leave the ground headed for their assigned targets, in the early years with B-36 aircraft it took days to prepare for and conduct combat operations. SAC's most-likely war scenario had aircraft taking off from their home station and flying to an AEC depot where their nuclear payload would be uploaded. They would then fly to their assigned forward operating location and await further orders. Once ordered into the air, the aircraft would drop their bombs on their assigned targets in the Soviet Union and those that survived the blast would land at alternate locations in Europe, the Middle East, or Japan.

On 12 September 1957, the 7th BMW at Carswell AFB, TX began requiring some B-36 crews to remain in base quarters in case of National Emergency during a test called Operation RED BALL in an apparent precursor to the SAC ground alert program, and to prove that the B-36 was still vital to national security. The B-36 also had considerable symbolic significance because, in the words of an Air Force general, it was "the first major weapon system to come into the operational inventory, superbly perform its deterrent mission, and be retired without firing a shot in anger."[7]

Firsts and Records

The B-36 was a truly incredible aircraft and unlike any other in its day. Because of its immense size and unprecedented wingspan, it could lift more weight to higher altitudes than any other aircraft. It was the first, and only, true intercontinental bomber capable of flights of over 10,000 miles without refueling and capable of remaining aloft for over 51 hours. Throughout its service life the Air Force conducted several challenging missions to prove B-36 capabilities. These flights were well-publicized and captured the American public's imagination. They were presumably also a great source of concern for the Soviet military.

On 8-9 April 1948, a B-36A (44-92013) made a long-range simulated tactical mission shuttling three times between Fort Worth to San Diego[1] during a flight that lasted 33 hours and 10 minutes covering 6,922 miles. The aircraft dropped 10,000 pounds of dummy bombs from 25,000 feet over the Air Force Bombing Range at Wilcox, AZ. The average speed was 223 mph but problems with two engines limited the speed to 214 mph, which was disappointing to Air Force leaders. The flight was conducted nonstop, and the aircraft had 399 gallons of fuel remaining when it landed in Fort Worth, which would have allowed the aircraft to cover 206 more miles.

On 13-14 May 1948, the same B-36A (44-92013) made another long-range simulated tactical mission shuttling from Fort Worth to San Diego[2] during a flight that lasted 36 hours and 8 minutes and covered 8,062 miles at an average speed of 223 mph. Gross weight was 299,619 pounds with 10,000 pounds of dummy bombs and 5,796 pounds simulated ammunition, plus ballasts to simulate guns not installed. The flight was conducted nonstop, and the aircraft had 986 gallons of fuel remaining when it landed in Fort Worth, which would have allowed the aircraft to cover 508 more miles. This flight, and the one conducted on 8-9 April, confirmed the B-36 was the first true intercontinental bomber.

On 18 May 1948, a B-36A dropped twenty-five 2,000-pound bombs (50,000 pounds total bombload) from 31,000 feet on the Naval Range at Corpus Christi, TX.[3] On 30 June 1948, another B-36A dropped 72,000 pounds of bombs during a flight test demonstrating the aircraft capability.[4] This was the maximum B-36A bombload and was the heaviest bombload ever carried by an aircraft at that time.

On 18 July 1948, a B-36A (44-92013) made another long-distance flight that covered 5,983 miles. The average speed was 301 mph, and this was the first aircraft to achieve this distance and speed. The gross weight was 310,292 pounds with 15,000 pounds of dummy bombs and a 3,555-gallon fuel tank in the bomb bay, plus ballasts to simulate guns not installed. The dummy bombs were dropped 9 hours and 49 minutes into the

B-36 "Fun Facts"

The XB-36 required the largest aircraft tire ever made; it weighed 1,475 pounds and used enough rubber for 60 car tires.

The B-36 was the only aircraft ever built that could carry two 43,000-pound bombs; this 86,000-pound bombload is more than a fully loaded B-24 Liberator.

The B-36 was the first, and only, true intercontinental bomber capable of flights of over 10,000 miles without refueling and capable of remaining aloft for over 51 hours.

At least one pilot commented the B-36 was so big it was like sitting on your couch and flying your house around.

One RB-36 was used to test the Boston Camera that could resolve a golf ball at an altitude of 45,000 feet and a slant range of 63,600 feet.

The XB-36 landing gear used the largest aircraft tire ever made and could only land at three bases. (U.S. Air Force)

The six radial engines on the B-36 had 28-cylinders containing 56 spark plugs each for a total of 336 spark plugs per aircraft.

The B-36 wingspan of 230 feet is longer than the first Wright Brothers flight in 1903; it still holds the record for the longest wingspan of any American combat aircraft.

The B-36 wing roots were 7.5 feet think and had tunnels that allowed mechanics to enter the wings in flight; total wing area was 4,772 square feet.

The B-36 could hold over 21,000 gallons of fuel in its wing tanks; enough for a car to circle the globe ten or more times.

At high speeds the B-36's ten engines developed more than 44,000 horsepower, roughly comparable to that of nine locomotives.

The volume of the B-36 bomb bay, 12,300 cubic feet, was equivalent to the capacity of three railroad freight cars; total volume of a B-36, nearly 18,000 cubic feet, is equal to three average five-room houses.

A B-36's deicing system generated 4,920,000 British Thermal Units of heat in one hour and could have heated a 600-room hotel or 120 five-room houses.

The B-36 electrical system contained 27 miles of wire; enough to wire 280 five-room houses.

flight at 2,811 miles. The flight was conducted nonstop, and the aircraft had approximately 1,528 gallons (5%) of reserve fuel remaining when it landed in Fort Worth, which would have allowed the aircraft to cover about 780 more miles.

On 13 August 1948, B-36A (44-92016) made a successful "shake-down" flight on Friday the 13th.[5] On 22 August 1948, a B-36A (44-92007) of the 7th BG made the first long-range flight with a crew comprised completely of Air Force personnel. The flight covered 15,500 miles in 26 hours and used 14,000 gallons of fuel. The crew said the aircraft was more comfortable than the B-29.[6]

On 12 March 1949, a B-36B (44-92035) from Carswell's 7th BG set a long-distance record for a 9,600-mile flight in 43 hours 37 minutes without refueling.[7] The aircraft took off on 10 March and stayed aloft for two days without refueling. It dropped a 10,000-pound simulated bombload in the Gulf of Mexico at 5,000 miles into the flight. The flight left Carswell and flew over Minneapolis, Great Falls, Key West, Gulf of Mexico, Houston, Fort Worth, Denver, Great Falls, Spokane, and back to Fort Worth proving its 10,000-mile range.

On 24 May 1950, the first landing of a B-36 outside North America occurred when four B-36 aircraft landed at Ramey AFB, Puerto Rico. Brigadier General C.S. Irvine, 7th BMW commander, flew in the lead aircraft.

On 12 September 1950, an 11th BG B-36D (49-2653) took part in the first D-model gunnery mission. This was the first B-36D assigned to Carswell AFB, TX. The test and evaluation mission was flown over the Eglin AFB, FL Gunnery Range at 24,000 feet. The aircraft experienced seven gunnery system failures of various types during the mission.

On 14 January 1952, an RB-36D (44-92090) took-off at 9:05 am from Fort Worth with a Convair crew on board and landed on 16 January at 12:35 pm, covering more than 10,000 miles in 51.5 hours. This was the longest known B-36 flight and was accomplished without refueling. Although this flight was unusual, most B-36 flights lasted at least ten hours and some lasted 30 hours. A typical training mission was scheduled for 24 hours.[8]

Operations

During 1949 SAC had only about 40 B-36 aircraft and only five to eight were considered operational.[1] The Air Force placed top priority on manning and equipping SAC to support the emphasis on the strategic air warfare mission by the JCS. In his first full year as SAC Commanding General[2], Lieutenant General Curtis LeMay began to make his lasting imprint on the command. "We didn't have one crew, not one crew, in the entire command who could do a professional job," LeMay wrote of the SAC he inherited. He challenged his crews to stage a practice bomb raid on Dayton, Ohio, from 30,000 feet, using photographs taken in 1941 – the best simulation they would have for the Soviet Union. After the fiasco that ensued, LeMay whipped the crews into shape. He moved the best people from other groups to make the nuclear capable 509th BG combat-ready, then did the same for the next most promising group.[3]

Training of bomber crews was intensified and accuracy of high altitude bombing substantially improved. Combat crew proficiency was raised through a system of lead crew training that proved successful during WWII. Units were deployed on a rotational schedule at overseas bases for limited periods to familiarize personnel with operating conditions outside the United States.[4] SAC continued to play an important role in the field of atomic energy through the development of strategies, tactics, techniques, and logistics to assure the most effective combat employment of atomic weapons in the national interest. Secretary Symington pointed out at the end of 1949: "Existence of this strategic atomic striking force is the greatest deterrent in the world today to the start of another global war."[5]

At 4:00 am local time on 25 June 1950, North Korean troops stormed across the 38th parallel. In November they were joined by Chinese "volunteers." These developments marked the end of President Truman's defense economy drive. First Germany, then Japan, then Russia, and now events in Korea had succeeded in advancing the cause of the B-36. Suddenly plenty of money was available for megabombers, and for supercarriers as well. The Korean War produced another milestone for SAC: Truman released nine atomic bombs to the military. They probably didn't leave the country, but the B-36 did, flying from Texas to airfields in Britain and Morocco in the spring and fall of 1951. Only six airplanes were involved, and their visits were short, but the message could not have escaped Moscow's attention. However briefly, the capital and most of the territory of the Soviet Union had come within the combat radius of the B-36.[6]

Operational Missions

When LeMay took command of SAC in 1948 he instilled a philosophy of realistic training and standardization across the command. He realized the training scenarios the crews were flying before his arrival resulted in scores so good, they were unbelievable. These training scenarios included bomb runs flown at 12,000 to 15,000 feet altitude, which was substantially below combat altitudes and allowed crews to operate without oxygen,

Maintenance Operations and Improvements

Like most new aircraft, the B-36 initially had its share of growing pains. After several improvements, the aircraft eventually was relatively trouble-free beginning with the B-36H and Featherweight versions.

The B-36B inherited many of the problems encountered with the B-36A. However, it introduced some new problems into the fleet. The nose turret was missing like the B-36A, but it never worked properly and was of little use anyway. The new R-4360-41 engines required extra fuel. New bomb bay tanks were supposed to be self-sealing but leaked throughout the aircraft service life. As with most new aircraft, parts shortages were a constant issue and parts cannibalization became a common practice. Shortages of B-36 specific support equipment left maintenance crews to rely on excess B-29 equipment.

Personnel and equipment required to get and keep a B-36 in the air. (U.S. Air Force)

Fuel leaks were a constant issue in early B-36 models that were not satisfactorily solved until the B-36H. Unreliable electrical systems were a concern through 1949 and resulted in system changes. Engine troubles were still an issue in 1950. Engine malfunctions at altitude, that could not be detected on the ground, were a major concern and required immediate installation of new airborne ignition analyzers.

The original propellers on all B-36 models were restricted until replaced with newly manufactured blades. In 1952, the entire B-36 fleet was restricted to 25,000 feet due to a faulty bulkhead that caused an RB-36 accident at 33,000 feet. The restriction was lifted after deficient bulkheads were replaced. All aircraft, except the first 152, were grounded in March 1952 after a B-36F landing gear failed while parked on the ramp at Carswell, and two crashes were traced to defective landing gears. The grounding order was lifted after replacement of the landing gear pivot shaft.

Convair and the Air Force initiated the Standardized Aircraft Maintenance-Strategic Air Command program (Project SAM-SAC) in 1953 to address B-36 maintenance problems. All aircraft cycled through the Convair plant in Fort Worth in three phases for intensive maintenance and modifications. At least 25 aircraft were in Fort Worth simultaneously through 1957. This project was a huge success and resulted in significant reliability improvements across the fleet. The project also developed the Quick Engine Change kit, now used on most aircraft, to reduce engine replacement times and the Big-Kel program (devised by the San Antonio Air Materiel Area at Kelly AFB, TX) to replenish flyaway kits for oversees deployments.

and for radars to work properly that would not at combat altitude. These missions also used large radar reflecting targets in the middle of the ocean that could be easily detected. Beginning in January 1949, under LeMay's direction, SAC crews began flying more realistic missions, including simulated bomb runs over several American cities that resembled targets in the Soviet Union. This soon became routine for all SAC units and crews.

On 12 September 1948, twelve B-36A aircraft took part in an Air Force Day celebration.[7] Ten aircraft were flown by Air Force crews logging 152 hours and 45 minutes total, and two by Convair crews. In October 1948, a B-36A flew a long-distance flight from Fort

Worth to Muroc AFB, California.[8] It flew at 37,000 feet and averaged 330 mph, dropping 12,000 pounds of bombs (one would not release).

In November 1948, a B-36B flew a 2,000-mile flight from Fort Worth to conduct a low-level bomb run for a group of high-level Air Force officers at the Eglin AFB, FL air show.[9] The aircraft dropped two groups of bombs from 10,000 feet at a simulated ship off the coast and caused heavy damage. On 5 December 1948, a B-36B completed a 14-hour long-range mission of 4,275 miles at 40,000 feet with an average cruise speed of 303 mph.[10]

B-36A (44-92022) on a simulated low-level bomb run. (American Aviation Historical Society)

From 7-8 December 1948, a B-36B completed a round-trip simulated mission from Carswell AFB, TX to Hawaii covering 8,100 miles without landing and proved its intercontinental capability.[11] The aircraft flew undetected into Hawaiian airspace (just seven years after the Japanese Pearl Harbor raid) and on the return to Carswell dropped a 10,000-pound bombload in the ocean near Hawaii. It averaged 236 mph in 35 hours, but despite its long range, the aircraft's relatively low speed and operating weight (approximately 225,000 pounds) caused criticism of the B-36 program.[12] The Air Force subsequently authorized the addition of four jets to improve speed. Most B-36B aircraft received the jets and were converted to B-36D.

On 12 December 1948, a B-36B completed a 14-hour long-range mission of 4,275 miles at 40,000 feet with an average cruise speed of 319 mph.[13] On 15 January 1949, a five-ship B-36 formation flew over the United States Capital commemorating President Truman's inauguration, including several low-level runs.[14]

A five-ship B-36 formation flew over the United States Capital on 15 January 1949 commemorating President Truman's inauguration. (American Aviation Historical Society)

On 29 January 1949, a B-36B dropped two 43,000-pound bombs on a practice target over Muroc Dry Lake, CA (one from 35,000 feet and one from 41,000 feet) at a speed of 350 mph. The aircraft averaged 250 mph and covered 2,900 miles. The forward bomb was released first, and nose pitched up (dropping aft first would cause the aircraft to pitch down and become potentially unrecoverable).[15]

On 22 April 1949, two B-36B aircraft from the 7th BG at Carswell AFB, TX flew to Muroc, CA for an accelerated service test. The mission tested the suitability of dropped very large type bombs from a B-36 at 40,000 feet. The test lasted a month and involved several missions. The aircraft returned home on 10 June 1949.

On 3 September 1949, two B-36B aircraft from the 7th BG and one from the 11th BG at Carswell AFB, TX completed a flyover at the Cleveland Air Races in Cleveland, OH. They accomplished two more flyovers during the event, one on 4 September and one on 5 September. On 14 September, one 7th BG B-36B flew a navigational training flight from Carswell to Eielson AFB, AK. It conducted a radar bombing run on Stockton, CA and Geiger Field, WA enroute to Alaska. The aircraft returned to Carswell on 16 September.

B-36 aircraft flew many demonstration flights to prove the aircraft's capabilities. (American Aviation Historical Society)

On 21 August 1950, 18 B-36B aircraft from Carswell AFB, TX departed for Limestone AFB, ME for a simulated combat mission. Limestone served as the pre-strike staging base for a simulated bombing of St. Louis, MO. The strike force included nine B-36B aircraft from the 7th BG and nine from the 11th BG. On 23 August, 17 of the bombers launched out of Limestone while one was grounded for maintenance problems. The force successfully completed its bombing mission and all aircraft were recovered at Carswell on 24 August.

On 20 September 1950, three B-36D aircraft from the 7th BG at Carswell AFB, TX participated in an exact war plan profile mission consisting of a night attack on Fort Worth. It completed additional training by conducting a simulated bomb run over Birmingham, AL and conducted a live fire over the Eglin AFB, FL Gunnery Range.

On 14 October 1950, three 7th BG B-36D aircraft took off from Carswell AFB, TX for a special training mission to determine the people, equipment, and supplies needed for staging through bases other than the unit's home station. They conducted a simulated bomb run over Fort Worth and then another over Phoenix, AZ. They then completed a gunnery mission over the Pacific Ocean before landing at March AFB, CA. They flew to Castle AFB, CA on 16 October. On 17 October, they took part in a camera gunnery mission with F-84 fighters at 25,000 feet over Southern California and then completed simulated bomb runs over San Francisco, Sacramento, Phoenix, and Fort Worth before landing home at Carswell.

On 30 November 1950, eight B-36D aircraft deployed from Carswell AFB, TX to Limestone AFB, ME to test pre-strike staging facilities and evaluate combat crews on a profile mission. The force consisted of four 7th BG and four 11th BG aircraft. On 1 December, six of these aircraft conducted simulated bombing missions over Charleston, SC and Tallahassee, FL enroute to Ramey AFB, Puerto Rico. This marked only the second time Carswell crews had visited Puerto Rico. A welcome change, no doubt, from the usual deployments to Alaska.

In January 1951, a 28th SRW RB-36D flew 200 hours in one month. In March, the 28th SRW flew over 1,000 hours with its RB-36D fleet. On 6 April 1955, a DB-36H launched a guided missile (GAM-63) from 42,000 feet; the explosion took place 6 miles above Yucca Flat, NV and was the highest known nuclear blast at the time.[16]

On 12 April 1951, 12 B-36D and five B-36B aircraft from the 7th BMW at Carswell AFB, TX flew night bombing missions in the Indianapolis, IN area to determine the wing's capability to bomb complex industrial targets. The B-36D aircraft also attacked the secondary target of New York City, while the B-36B aircraft also attacked Oklahoma City and Austin, TX before flying the primary mission over Indianapolis. The aircraft all returned to Carswell on 13 April. The force repeated these missions again from 16-17 April.

On 10 July 1951, nine B-36 aircraft from the 7th BMW at Carswell AFB, TX took part in a special training mission including a high-altitude penetration of Eglin AFB, FL using F-84 fighter escorts. The aircraft flew from Carswell to Port Arthur, TX where they picked up their fighter escorts from the 12th Fighter Wing, Bergstrom AFB, TX. From there they

The 28th SRW flew over 1,000 hours with its RB-36D fleet in March 1951. (University of North Texas Libraries)

headed east towards Eglin. Several F-86 fighters from Eglin intercepted the bombers enroute. The bombers completed the mission and returned to Carswell.

On 4-9 August 1951, 15 B-36 aircraft from the 7th BMW at Carswell AFB, TX conducted a radar bombing evaluation against a commercial target at Binghampton, NY to compare radar bombing accuracy using eight-year-old photography against recent radar reconnaissance. All 15 aircraft launched out of Carswell on 4 August. Eight aircraft used the old photography and seven had the new. The aircraft with the new photography scored effectively and those with the old did not. On 9 August eight of the aircraft employed the new photography and scored effectively.

On 11 October 1951, the 7th BMW at Carswell AFB, TX conducted a simulated combat mission using three of its newly assigned B-36F aircraft. The mission was flown in the Eglin AFB, FL range and all aircraft completed the mission as scheduled. They returned to Carswell on 12 October.

On 12 July 1953, three B-36H aircraft flew simulated attacks during Operation TAIL-WIND on three vital control centers of the Air Defense Command at Savannah, GA, and New York City.

On 14 March 1954, 19 B-36 aircraft launched for Operation PATHAND on a unit simulated combat mission to Goose Bay, Labrador and flew strike missions on American cites while returning to base.

From 30 April to 5 May 1954, the 7th BMW and other SAC units participated in Operation ALAMO and conducted night simulated radar bombing missions on an industrial complex in San Antonio, TX.

On 9 July 1954, 22 B-36 aircraft conducted Operation CHECK POINT striking industrial targets in the northeastern United States and southeastern Canada before deploying to North Africa in August. From 11-22 October, 28 B-36 aircraft participated in Operation FAT CAT, a combined operational readiness and unit simulated combat mission.

On 3 March 1955, eight B-36 aircraft bombed Bedford, IN under Operation BAGDAD BILLY. On 6-7 October, 24 B-36 aircraft took part in an evacuation mission called Operation POST HOLE.

On 28 February 1956, three B-36 aircraft won first place in a competition against other SAC wings called Operation SNOWBANK. On 15 March 1956, 13 B-36 aircraft conducted simulated bomb runs on San Antonio, Houston, and Little Rock under Operation HORNET GULF.

On 21 March 1956, eight B-36 aircraft flew simulated bomb runs during Operation HORNET HOTEL on San Antonio, Houston, Springfield, MO, Denver, Salt Lake City, and Phoenix. F-80 fighters from Hensley AFB, TX (near Dallas) flew intercepts against the bombers.

From 6-8 June 1956, 11 B-36 aircraft participated in Operation HORNET JULIET and conducted simulated attacks on Canada and the eastern and southeastern United States. From 26-27 June 1956, B-36 crews took part in a secret mission called Operation BROAD JUMP.

On 27 November 1956, B-36 bombers flew simulated missions in the central United States called Operation HORNET MIKE. On 13 December, nine B-36 aircraft tested the capability to successfully launch aircraft under simulated wartime conditions called Operation HAPPY BIRTHDAY.

On 10 January 1957, 30 B-36 aircraft flew bomber stream missions called Operation WEDDING ALPHA. On 17 January they flew Operation WEDDING BRAVO and on 24 January they flew Operation WEDDING CHARLIE.

On 6 February 1957, 30 B-36 aircraft flew a simulated combat exercise called Operation FIRST TEAM. On 5 March, 27 B-36 aircraft flew a simulated combat exercise called Operation LAST STAND.

From 1 April to 1 July 1957, Project LONG RANGE trained crews in a new type of mission. On 5 April, six B-36 crews took part in a special weapon exercise called Operation BRIAR RABBIT.

On 1 July 1957, 13 B-36 aircraft completed a bomber stream mission called Operation LOGBOOK ECHO. On 18 July they completed Operation LOGBOOK DELTA. On 28 July, B-36 crews trained with new plans under Operation OVER EASY. On 12 September, 25 B-36 aircraft completed a simulated bomb mission called Operation TREASURE.

Operational Deployments

While it never dropped a bomb in anger, the very existence of the B-36 had a deterrent effect on America's enemies. Some think it may have prevented direct Soviet entry into the Korean War on the side of North Korea and China.[17] The B-36 was also deployed as a deterrent during the Suez Crisis of 1956 and the Hungarian revolt against Soviet occupation that same year. The Air Force exploited this deterrent effect and deployed the B-36 outside the United States on several occasions as a show of force and to test new strategic concepts for the atomic age.

On 17 January 1949, a B-36 left Eglin AFB, FL for an 18 hour and 15-minute flight to Ladd AFB, AK where the temperature was -38 degrees Fahrenheit. The aircraft remained in Alaska for two months where it underwent cold weather testing. Testing continued in Alaska from 14 December 1949 to 8 April 1950 when low temperatures reached -53 degrees Fahrenheit and "highs" reached -39 degrees Fahrenheit.[18] These tests were intended show that that B-36 could operate in these extreme temperatures, although the truth is they required significant preheating before they could get airborne.

Following the GEM modifications, 7th BMW crews routinely deployed B-36B aircraft to Goose Bay, Labrador and Eielson AFB, AK on a rotational basis. Under Operational Order 19-49 the 8th AF ordered the 7th BMW to initiate Operation DRIZZLE for training and pre-strike staging to gain Arctic experience. These GEM-modified B-36B aircraft were famously painted bright red on the wingtips and tail so they could be located during Arctic white-outs or if they were forced down in remote snow-covered terrain.[19]

On 16 January 1951, six B-36D aircraft from the 7th BMW at Carswell AFB, TX arrived at RAF Lakenheath, UK under Operation UNITED KINGDOM becoming the first B-36 aircraft deployed to England. The force included three aircraft from the 7th BG and three from the 11th BG. At total of 11 aircraft had launched from Carswell two days earlier for staging through Limestone AFB, ME. However, two aircraft aborted on take-off from

GEM-modified B-36B aircraft were painted bright red on the wingtips and tail so they could be located during Arctic white-outs or if they were forced down in remote snow-covered terrain. (U.S. Air Force)

Limestone on 16 January due to engine failures and three others returned to Carswell later that day. The remaining aircraft took-off on 16 January and completed a night bombing attack on Helgoland, Germany and a simulated bomb run on the Heston Bomb Plot in London before landing at Lakenheath. Crews flew missions out of Lakenheath for the next four days to evaluate the B-36D under simulated war plan conditions, evaluate airspeed and compression tactics for heavy bomber aircraft, and evaluate crew capability for bombing unfamiliar targets. They returned to Texas on 20 January.

On 19 February 1951, five B-36D aircraft from the 7th BMW at Carswell AFB, TX deployed to Eielson AFB, AK to take part in a mock bombing run over Portland, OR. The aircraft departed Eielson on 22 February and flew a high-level bomb run over Portland before returning to Carswell later that day.

On 17 July 1951, six B-36D aircraft from the 7th BMW at Carswell AFB, TX deployed to Goose Bay, Labrador to familiarize crews with the staging base, test its capabilities, and provide crews with actual polar experience. On 17 July, the aircraft flew a polar navigation training flight and recovered at Goose Bay. They departed on 23 July and completed a partial war plan profile including bomb runs over Tampa, Birmingham, Fort Worth, Memphis, Little Rock, and Dallas before landing at Carswell on 24 July.

On 16 September 1951, the 7th BMW at Carswell AFB, TX deployed another six B-36D aircraft to Goose Bay, Labrador to test the staging base's capabilities and facilities. While there, the crews flew a polar navigation mission to Thule AB, Greenland. The aircraft departed for Carswell on 23 September and completed a partial profile mission enroute to home station.

On 11 October 1951, the 7th BMW deployed for its final familiarization flight to Goose Bay. Again, they completed a polar navigation mission to Thule and conducted a partial profile mission enroute to home station on 17 October. The wing now had all squadrons capable of staging out of Goose Bay and was ready to deploy at a moment's notice.

On 3 December 1951, six 11th BMW B-36D aircraft took off from Carswell AFB, TX flying nonstop and arriving at Sidi Slimane AB, French Morocco as the first B-36 aircraft deployed to Morocco. They returned to Texas on 6 December.[20]

From August to September 1953, the 92nd BMW at Fairchild AFB, WA made the first mass B-36 flight to the Far East (Japan, Okinawa, and Guam) during Operation BIG

92nd BMW B-36D aircraft dispersed to Yokota AB, Japan and other bases in the Far East during Operation BIG STICK. (American Aviation Historical Society)

STICK. This was a 30-day exercise after the Korean War to show United States determination to maintain peace.[21] The wing deployed 20 B-36 aircraft to Kadena AB, Okinawa to test the capabilities of the B-36 in long distance flight. The operation exercised SAC's Emergency War Plan by conducting bombing training in Alaska and the Far East. It also assessed the operational suitability of overseas installations such as Kadena, Yokota AB, Japan, and Andersen AFB, Guam for heavy bombers. In late August, 15 of the 20 aircraft flew from Fairchild to Eielson AFB, AK and prepositioned for simulated strikes on targets in the North Pacific before landing at Kadena. The remaining five aircraft were Featherweights, and they flew a 6,500-mile great circle route directly to Kadena. Although well-planned, weather enroute forced the groups to break up resulting in several cells of aircraft arriving simultaneously at Kadena from different directions. There were some close calls, but all aircraft landed safely. The aircraft flew several training missions throughout the deployment, landing at Yokota and Anderson to survey the installations. The aircraft began returning home on 29 August and all but three were home by early September. On 5 September, two aircraft flew from Kadena and conducted a fly-over at the Dayton, OH airshow. Impressively, they met a control time rendezvous with six B-47 bombers coming from the UK. These aircraft then landed at Wright-Patterson and Lockbourne AFB, OH before arriving at Fairchild on 9 September. The final B-36 arrived at Fairchild on 12 September after maintenance problems. The 92nd BMW received the Air Force Outstanding Unit Award on 15 May 1955 for this effort, as well as high praise from General LeMay.[22]

On 15-16 October 1953, the 92nd BMW at Fairchild AFB, WA deployed for the first 90-day rotational training assignment to Andersen AFB, Guam. It was the first time an entire B-36 wing deployed outside of the United States.[23] The 92nd BMW still had B-36D aircraft prone to fuel leaks, especially in the winter weather of Fairchild and the northern bases to which they usually deployed. The B-36H and B-36J had better sealant and fewer leaks.

On 2-3 February 1953, 18 B-36 aircraft from the 7th BMW left Carswell enroute to their staging base at Goose Bay, Labrador under Operation STYLESHOW. One aircraft developed mechanical problems and returned to Carswell. The remaining 17 aircraft left Goose Bay on 6 February heading for their deployment location at RAF Fairford, UK and encountered severe weather enroute. One B-36H (51-5719) ran out of fuel after excessive holding in the pattern and the crew abandoned the aircraft with no injuries. All but two aircraft, held back for special weapons training, departed Fairford for Goose Bay on 13 February. One B-36H (51-5729) was misdirected by Goose Bay controllers and flew into a hill killing two of the 17 crew onboard. The remaining 13 aircraft arrived back at Carswell on 21 February (the two that remained at Fairford arrived at Carswell on 23 February).[24]

In March 1954, two RB-36 aircraft from the 5th SRW at Travis AFB, CA deployed to Thule AB, Greenland for a two-week period. Each aircraft flew at least two missions in support of Project 54AFR-11 along the northeastern coast of Greenland and near Soviet territory to monitor Soviet transmissions. On the first mission each aircraft monitored the area of interest for 12 hours giving 24 hours of total coverage. This mission was considered "highly successful" but actual results are uncertain. On the second mission they also conducted experimental flash photography over the highly reflective snow-covered terrain of Greenland.[25]

From 13-16 September 1954, 6th BMW B-36F aircraft departed Walker AFB, NM for a 90-day deployment to Andersen AFB, Guam as part of a continuing rotation of B-36 wings. Aircraft flew from Walker to Phoenix, AZ and then over San Diego, CA before making the 2,620-mile leg to Honolulu, HI and then a ten-and-a-half-hour leg to Wake Island, before completing the final 1,500 mile, six-and-a-half-hour leg to Andersen. The total flight time was 32 hours non-stop and 7,150 miles without refueling. During the

B-36F aircraft of the 6th BMW enroute from Walker AFB, NM for a 90-day deployment to Andersen AFB, Guam as part of a continuing rotation of B-36 wings. (National Museum of the U.S. Air Force)

deployment, 6th BMW aircraft made a no-notice flyover for an airshow at Manila Airport in the Philippine Islands and a low-profile 3,400-mile, 15.5-hour, cargo run to Bangkok, Thailand to deliver arms and munitions for covert American troops operating in Southeast Asia. The wing began redeploying to Walker on 7 December and continued through the end of the month. Their replacement wing began arriving on Guam the second week of January 1955.

In March 1955, an 11th BMW B-36 made the 25-hour non-stop flight from Carswell to Buenos Aires to "buzz" the inauguration ceremony of Argentinian President Luis Batlle Berres at his request. The aircraft flew over Mexico City, Cocos Island, Lima, and the 23,000-foot Andes Mountains marking the first and only trip of a B-36 to South America.

On 20 July 1955, 20 Carswell B-36 aircraft flew out of Nouasseur AB, French Morocco for Operation DEEP ROCK and conducted a test and evaluation mission for SAC against targets in the northeast and eastern United States.

From 13-14 September 1955, B-36 aircraft conducted a unit simulated combat mission called Operation PEPPER POT II on their return from Nouasseur AB, French Morocco.

From 14-28 February 1956, 25 B-36 aircraft deployed to Nouasseur AB, French Morocco after loading special weapons at Loring AFB, ME under Operation STYLESHOW.

Competitions and Exercises
In 1948, General Kenney decided to hold the first SAC Bombing Competition, typically called Bomb Comp, hoping to stimulate interest in improving bombing accuracy. The competition was held from 20-27 June at Castle AFB, CA. Ten B-29 groups participated but the B-36 was not yet in SAC's inventory. Each crew accomplished three visual and three radar releases from 25,000 feet. The 43rd BG was the top unit and received the trophy.[26]

The second SAC Bomb Comp was held from 3-7 October 1949, with twelve bomb groups participating including three B-36, seven B-29, and two B-50. A B-36 crew from the 28th BG at Rapid City AFB, SD won the individual aircrew trophy. The 28th BG winning crew average visual bombing error of 441 feet and radar bombing error of 1,053 feet raised doubt about accuracy and ability to conduct precision strikes with conventional bombs. The AN/APQ-24 bomb system was not as accurate or reliable as expected.[27]

No SAC Bomb Comp was held in 1951 due to the demands of Korean War operations. The third SAC Bomb Comp was held from 13-18 August 1952. A navigation competition was included for the first time to allow reconnaissance wings to compete. Twelve SAC bomb wings and three reconnaissance wings participated flying B-36, B-29, and B-50 aircraft. The B-36 aircraft staged out of Carswell AFB, TX while the remaining aircraft staged out of MacDill AFB, FL. Two RAF crews flying B-29 aircraft also participated.[28]

The Fairchild Trophy was presented for the first time and would become the coveted trophy of future competitions. The 97th BMW, a B-50 unit, won the trophy.

From 12-15 December 1951, six SAC aircraft and crews participated in the first RAF Bomber Command bombing competition. The aircraft operated out of RAF Sculthorpe, UK. SAC competitors included one B-29 from the 9th BMW (Travis AFB, CA), one B-29 from the 301st BMW (Barksdale AFB, LA), two B-36 aircraft from the 7th and 11th BMW (Carswell AFB, TX), and two B-50 aircraft from the 93rd BMW (Castle AFB, CA). The SAC B-29 team (9th and 301st BMW) placed first in the competition.[29]

SAC's fourth Bomb Comp was held from 13-18 October 1952 and included ten B-29, five B-50, and four B-36 wings. The B-36 aircraft staged out of Walker AFB, NM and the remaining aircraft staged out of Davis-Monthan AFB, AZ. The RAF also entered two B-29 aircraft. The 97th BMW and the 93rd BMW tied for the Fairchild Trophy. General Thomas Power, Vice Commander of SAC, flipped a coin to determine the winner. The 93rd BMW walked away with the honor. From 23 October - 1 November 1952, the first SAC Reconnaissance Competition was held. The 28th SRW (Rapid City AFB, SD), an RB-36 unit, won the competition and P.T. Cullen award.[30]

The 28th SRW, an RB-36 unit, won the competition and P.T. Cullen award in 1952 during the first SAC Reconnaissance Competition. (American Aviation Historical Society)

From 18-27 October 1953, the second SAC Reconnaissance Competition was held. The 5th SRW (Travis AFB, CA), an RB-36 unit, won the competition and P.T. Cullen award. SAC continued its Bomb Comp by hosting the fifth competition from 25-31 October 1953. The B-47 made its first appearance, and seven wings staged out of Davis-Monthan AFB, AZ. Four B-36, four B-50, and two B-29 wings staged out of Walker AFB, NM. The 92nd BMW (Fairchild AFB, WA), a B-36 unit, barely edged out a B-50D wing to win the Fairchild Trophy.[31]

From 9-14 August 1954, RB-36 units continued to dominate the SAC Reconnaissance Competition. The 28th SRW won the P.T. Cullen Award. The sixth SAC Bomb Comp was held from 23-29 August 1954. It included fifteen B-47 wings staging out of Barksdale AFB, LA. Six B-36 and two B-50 wings staged out of Walker AFB, NM. One 28th SRW RB-36 crew competed due to its outstanding performance in the reconnaissance competition. B-36 wings continued to dominate and won the top three positions in bombing and navigation. The 11th BMW (Carswell AFB, TX) won the Fairchild Trophy.[32]

B-36 units won the coveted Fairchild Trophy during SAC Bomb Comps in 1953, 1954, and 1956. (American Aviation Historical Society)

From 24-30 August 1955, the seventh SAC Bomb Comp was held. It included only B-47 and B/RB-36 aircraft since the B-50 had been phased out. The B-47 aircraft staged out of March AFB, CA and the B/RB-36 aircraft, now classified as dual role bomber and reconnaissance aircraft, staged out of Fairchild AFB, WA. For the first time a B-47 unit, the 320th BMW (Mather AFB, CA), beat out the B/RB-36 units and won the Fairchild Trophy in the SAC Bomb Comp. Since all RB-36 aircraft were converted to bombers, the SAC Reconnaissance Competition held from 24-30 Sep 1955, became an all RB-47 event. The 91st SRW won the P.T. Cullen Award.[33]

From 24-30 Aug 1956, for the first time, the SAC Bomb and Reconnaissance Competition were a combined event. Recon aircraft competed against the bombers in bombing and navigation, and against each other in recon. It was the largest such event ever held with 42 wings participating. It included B-52 aircraft for the first time, along with B-47, RB-47, and B/RB-36 aircraft. The 11th BMW (Carswell AFB, TX), a B/RB-36 unit, beat out the B-47 and B-52 units to win the Fairchild Trophy. The 91st SRW, an RB-47 unit, won the P.T. Cullen Award.[34]

Except for crew and wing navigation awards, B-47 units beat out the B/RB-36 units in all events of the SAC Bomb Comp held from 30 Oct - 6 Nov 1957. The 321st BMW (McCoy AFB, FL), a B-47 unit, won the Fairchild Trophy.[35]

The final appearance of the B/RB-36 in a SAC Bomb Comp occurred from 13-18 Oct 1958. B-47 units dominated the competition and the 306th BMW won the Fairchild Trophy.[36]

Atomic Tests

Between 1952 and 1956, various Air Force aircraft supported atomic tests including the B-36, B-47, and B-50. B-36 aircraft participated in atomic weapons tests under several operational code names. Sometimes they dropped bombs. Other times they were used for "effects aircraft" to determine the effect of high-yield nuclear bombs (heat, blast, and overpressure) on aircraft in flight. Other times they were used as "sampler aircraft" to measure radiation released from nuclear bombs. The B-36 was used primarily because it was still SAC's front-line bomber, and it could carry large amounts of equipment over long distances. Another major consideration was the potential effect of the blast on fabric control surfaces. The B-50 still used fabric control surfaces. Although the original B-36 design used fabric control surfaces, including the rudder, these were later replaced with metal control surfaces making the aircraft less susceptible to heat than the B-50.

Four EB-36H aircraft (51-5726, 51-5731, 52-1357, and 52-1358), later redesignated JB-36H, were assigned to the 4925th Test Group (Atomic) at Kirtland AFB, NM to support atomic test operations. Two JRB-36H (51-5748 and 51-5750) aircraft were assigned to the 4925th and modified with an upward looking camera to photograph mushroom clouds. The 4925th also used two B-36H (50-1083 and 50-1086) and one RB-36D (52-1386) aircraft as sampler aircraft.

One B-36D (49-2653) was used extensively as an effects aircraft and eventually received structural damage beyond repair. It was modified at Wright-Patterson AFB, OH for atomic testing with equipment including thermocouples, accelerometers, strain gages, meters, and oscilloscopes. Data was recorded on aircraft using photos of oscilloscopes and meters or "wire recorders" since telemetry was not yet available, and Electro-Magnetic Pulse (EMP) energy probably would have eliminated its use anyway. The aircraft sported "Ruptured Duck" nose art and was painted with white heat resisting paint on bottom (before this was standardized across the fleet).[37]

Operation TEXAN. The Air Force conducted Operation TEXAN on 13 August 1952 as a rehearsal for Operation IVY. A United States Navy ship was positioned in the Gulf of Mexico 425 miles from Bergstrom AFB, TX. This was the same distance as the Eniwetok test site was from Kwajalein Atoll. The operation involved 39 aircraft including the B-36D (49-2653) effects aircraft and two B-36H drop aircraft (one of these also served as drop aircraft and one as sampler aircraft during Operation IVY). The device dropped was a modified T-59 training shape (really an Mk-4) that simulated the Mk-18 dropped at Eniwetok. The device was 60 inches in diameter, 128 inches long, and weighed 8,600 pounds.[38]

Operation IVY. This operation involved two shots designed to test the effects of megaton yield atomic weapons. Shot Mike (Megaton), tested a 10.4 megaton liquid-fueled thermonuclear device on 31 October 1952 (1 November local). The bomb was 1,000 times more powerful than the bomb dropped on Hiroshima. It was too heavy to be air-dropped and was detonated in a tower on the ground. The blast was the largest ever seen up to that time and created a 164-foot-deep crater, that was 6,240 feet in diameter. The fireball was 3.5 miles in diameter. The B-36D (49-2653) effects aircraft was 15 miles out and 40,000 feet high. When the shock reached the aircraft, it was at 38,500 feet and 25 miles slant range. It experienced 93 degrees Fahrenheit on the wing tip nearest the blast and 0.33 psi over pressure (only 62% of design limit and less than expected). The conclusion from the test was that the B-36 was not fast enough to escape the blast, and parachutes would be required for air dropped atomic bombs. SAC initially opposed it but then agreed to add them for almost all nuclear weapons.[39]

On 7 November 1952, a prototype Mk-18 fission device arrived on Kwajalein for Operation IVY, Shot King (Kiloton). The Super Oralloy Bomb (SOB) was intended to be air dropped and consisted of standard stockpile components with 132 pounds of highly enriched uranium (Oralloy) and weighed 8,600 pounds. A dry run was conducted on 8 November when a T-59 training device was dropped in the ocean near the planned target. The SOB was loaded on the B-36H drop aircraft on 12 November and the aircraft

All B-36 aircraft were eventually painted with white heat-resistant paint because of the extreme temperatures experienced during Operation IVY testing. (American Aviation Historical Society)

took off on the morning of the 13th. However, bad weather over the target forced a mission abort and the aircraft returned to station with the bomb still loaded. It was then downloaded and underwent minor maintenance.

On 16 November, the bomb was reloaded, and the drop aircraft took off at 0630 local time. It released the bomb at 40,000 feet after three hours of practice runs. The bomb exploded at 1,480 feet (500 kiloton yield) 56 seconds after being released. The B-36D (49-2653) effects aircraft was flying at 28,000 feet off target and little useful data was obtained. The B-36H drop aircraft sustained minor damage and engineers calculated it was exposed to 80% of design limit. New thermal protection devices (thermal curtains) were developed and placed in the cockpit windows to protect the crew from the blast. Also, this test led to eventual painting of the bottom of all B-36 (and other SAC aircraft) with white heat-resistant paint (various paints and schemes were tried during the remaining tests).

Operation UPSHOT-KNOTHOLE. This operation was conducted from 17 March - 4 June 1953 and included 11 tests (3 airdrop, 7 tower, and 1 airburst). The tests were conducted at the Nevada Test Site. The purpose was to validate new theories using both fusion and fission devices.[40] Shot Nancy was conducted on 24 March 1953 using a tower launched 24 kiloton weapon. A fleet of 53 aircraft participated including twelve B-36 aircraft. The pre-dawn explosion was conducted to provide first-hand knowledge of what an atomic explosion looked and felt like.[41] All B-36 aircraft from various units landed at Carswell.

The B-36D (49-2653) effects aircraft participated in Shot Encore on 8 May 1953. It had previously returned to Wright-Patterson on 21 November 1952 after Operation IVY, Shot King, failed to provide adequate data. The aircraft was modified with new instrumentation but was not properly calibrated before flying to Kirtland AFB to participate in this test. The 27-kiloton weapon was airdropped, and the effects aircraft was loaded with a 25,000-pound dummy load to simulate an operational aircraft. It flew widely separated from the drop aircraft and the data was still insufficient because the instrumentation was not calibrated. It was sent to Carswell for calibration (by Cook research Labs) after the shot was completed.

On 4 June 1953, a B-36H dropped a 61-kiloton Mk-7 weapon for Shot Climax over the Nevada Proving Grounds. Since the B-36 was not normally equipped to handle the tactical Mk-7, a bomb rack from an F-84 fighter was temporarily installed on the B-36 for the test. The weapon was airburst at 1,334 feet and produced a yield of 61-kiloton which was the highest of any United States continental test up to that time.[42]

Operation CASTLE. This operation was conducted with six shots from 1 March to 14 May 1954. All shots produced megaton range yields. The B-36D (49-2653) effects aircraft participated in all six shots. For the first five shots the aircraft was flying away from the blast and on the last shot it was flying toward the blast. Sufficient data was obtained from five of the shots, but the third (Shot Koon) on 7 April failed to produce a high enough yield to obtain useful data.

Shot Bravo, on 1 March, was the largest nuclear weapons test ever conducted by the United States which produced a 15-megaton yield and the overpressures were greater than 100% of the aircraft design load. The blast caused mechanical damage that required replacement of the bomb bay doors, all four lower plexiglass scanners' blisters, and the radar antenna radome. The undersides of all six engine nacelles and the landing gear and lower aft gun turret bay doors were also damaged. The blast caused considerable heat damage on the underside of the aircraft, including buckled inspection panels and blistered paint on the wing roots, horizontal stabilizers, elevator trim tabs, and many other components. The crew saw a red glow through the thermal curtains from the blast that lasted about 15 seconds. The aft crew cabin filled with smoke, the engine fire warning lights illuminated for about four seconds from the thermal radiation, and the Exhaust Gas Temperature (EGT) of the four jet engines pegged momentarily and then returned to normal.

Mushroom cloud from Operation CASTLE, Shot Bravo, which was the largest nuclear weapons test ever conducted by the United States. (National Oceanic and Atmospheric Administration)

Shot Yankee, on 5 May, produced 13.5 megatons and subjected the aircraft to 64% thermal loading and 76% gust loading. The 322 degrees Fahrenheit thermal pulse buckled the elevator skin and blistered the white thermal paint on the bottom of the fuselage. The aircraft was damaged beyond repair and written off on 27 June 1955.

Two B-36H (50-1083 and 50-1086) sampling aircraft also participated in Operation CASTLE. An RB-36H (52-1386) was used as a sampler controller aircraft that directed the sampler aircraft to the correct area of the clouds. The sampling aircraft reached 55,000 feet and often took two hours for the climb. The crew wore pressure suits and was limited to 90-100 minutes above 50,000 feet.[43]

Information gathered from Operation CASTLE resulted in a B-36 delivery handbook being developed. The preferred delivery flight began at 345 knots at 40,000 feet followed by a violent turn away from the target immediately after the bomb was dropped. It was determined the aircraft could survive a surface blast from a 10.8 megaton free fall or 100 megaton parachute-retarded bomb. If the bomb was set to explode at 6,000 feet these figures dropped to 1.5 and 3.5 megatons, respectively.[44]

An RB-36H (52-1386) was used as a sampler controller aircraft that directed the sampler aircraft to the correct area of the clouds. (American Aviation Historical Society)

Operation TEAPOT. Several operational B-36 crews from the 7th BMW at Carswell AFB, TX participated in training missions to verify the nuclear bomb delivery and escape procedures from the SAC weapons delivery handbook. The 45-kiloton Shot Turk occurred on 7 March 1955 with the B-36 aircraft one mile east of ground zero, flying eastbound at 250 mph and 23,000 feet. The 24-kiloton Shot Met occurred on 15 April 1955 with the B-36 aircraft directly overhead at 26,000 feet and 250 mph. Radiation exposure was essentially nil.[45]

A B-36 dropped a Mk-12 case to test a 1.2-kiloton weapon using the Range Able uranium core in a new compact, lightweight implosion system for Shot Wasp at the Nevada Proving Grounds on 18 February 1955. The drop was delayed until 12 noon due to clouds over the test location. The total weight of the bomb was 1,500 pounds and was the lightest nuclear explosive system fired up until this time. A B-36 dropped a non-nuclear high-explosive device from 30,000 feet over Yucca Flat, NV on 25 March 1955 in preparation for the nuclear test scheduled for 29 March. The device was used to provide photo reference points. Shot Wasp Prime occurred on 29 March 1955 and was largely a repeat of the previous Shot Wasp. About 40 Air Force aircraft participated in the test, mostly performing cloud sampling missions.[46]

A B-36H dropped a 3.2-kiloton air-to-air missile warhead like the Shot Wasp device for Shot HA (High Altitude) on 6 April 1955 over Yucca Flat, NV. It was dropped from 42,000 feet and the device was parachute retarded marking the first time this was done at the Nevada Proving Ground. The device detonated at 32,650 feet which was the highest altitude of any nuclear blast at that time. The B-36H dropped radiosondes to collect pressure readings and two RB-36 aircraft collected radioscope samples (one could not climb high enough to obtain useful samples).[47]

Operation REDWING. The first air drop of a deliverable hydrogen bomb occurred on 20 May 1956 at Eniwetok during Shot Cherokee. A B-36 was planned as the drop aircraft, but analysis indicated it would be too slow and a B-52B (52-013) was used instead. However, the B-36 dropped diagnostic cannisters for the shot. A B-36 dropped a low-yield TX-28 device for Shot Osage on 17 June 1956. This was the proof test for the W-25 lightweight, low yield, plutonium warhead intended for air defense and other tactical applications. This was the last United States nuclear test in which a B-36 participated.[48]

Operation MIAMI MOON. Four RB-36 aircraft participated in the British nuclear weapons tests Operation GRAPPLE on Malden Island (south of Hawaii). The aircraft were specially modified with external sampling pods to trap radioactive particles. A 17-hour calibration mission was flown on 11 May 1957, and two missions were flown on 15 May, two on 31 May, and two on 19 June. All aircraft launched from Hickam AFB, HI and the mission lasted 19 hours with about 2.5 hours inside the mushroom cloud at 40,000 feet.[49]

Operation OLD GOLD. During late 1957 and early 1958 two RB-36 aircraft from the 5th BMW deployed to Eielson AFB, AK to gather data from Soviet nuclear tests. The aircraft were fitted with two sampling pods that collected airstream air and ducted it into cannisters in the camera compartment. The flights were flown along the edges of Soviet airspace for several days following each Soviet test.[50]

COSMIC RAYS. Although not part of the atomic test series, one JB-36H (51-5726) from the 4925th Test Group (Atomic) at Kirtland AFB, NM was used during an investigation into the number and energy of cosmic ray neutrons in the upper atmosphere. The flights occurred between November 1956 and 1 February 1957. They were considered very successful and later continued with a B-52.[51]

Texas Twister
On Labor Day, 1 September 1952, a skeleton crew worked the B-36 flightline at Carswell AFB, TX. Most base personnel were picnicking with families at Lake Worth next

to the base or at other Labor Day events. The day was hot and humid, and forecasters predicted late day thunderstorms with winds up to 60 mph, nothing unusual for a late summer day in Texas. As a standard precaution against the potential high winds the flightline crews secured the aircraft with three-eighths inch steel tie downs.

A severe thunderstorm rolled into the area around dinnertime. At 6:42 p.m., base personnel were convinced a tornado was in progress. The madly spinning anemometer in the control tower recorded winds of 90 mph before the device broke. Troops took cover in concrete or brick buildings. Winds swept across the flightline, with enough force to pick up aircraft, rip them from their steel cable moorings, and fling them into each other, into hangars, or other objects. Flying debris acted like missiles, ripping through aircraft skins, and giving the impression afterward that the bombers had been repeatedly raked by cannon fire. Cockpits were crushed and access doors ripped off. The wind blew big chunks off the roofs of many buildings and tore great gaping holes in hangars. Had winds been under the 60-mph forecasted that day, the steel cables probably would have been enough to hold the aircraft down, but wind speeds were likely well above the recorded 90 mph, creating "takeoff" conditions for some of the aircraft. When it was over, the flight line was a tangle of airplanes, equipment, and pieces of buildings.[52]

The Labor Day tornado damaged 82 aircraft at Carswell in just a few seconds. Some were more severely damaged than the others, but all required significant repairs before they could be returned to service. (U.S. Air Force)

The tornado[53] damaged 82 aircraft in just a few seconds. Some were more severely damaged than the others, but all required significant repairs before they could be returned to service. Faced with the loss of a significant portion of the nation's nuclear strike capability and forced to declare the 19th AD non-operational, Lemay directed his staff to devise a plan to quickly return the aircraft to service.

By daybreak the next morning, LeMay and his staff had come up with a response plan called Project FIXIT. SAC personnel would perform repairs on the least-damaged third of the bombers, AMC personnel from Kelly AFB, TX, where the B-36 depot was located, would be responsible for another third, and the most heavily damaged remainder would be towed across the field to the Convair factory. The effort began immediately, with crews working around the clock. SAC put the two Carswell wings on an 84-hour workweek until the fixes were complete. The first B-36 was returned to service within a week of the storm and nine more were put back in service the week after that. Just one month after the disaster, 51 bombers had been returned to service and the two Carswell wings were once again declared operational. It took until 11 May 1953 to get the last storm-damaged B-36 back on flying status.[54]

Only two of the 82 damaged aircraft were not returned to service. B-36D (44-92051) was shipped to Kirtland AFB, NM where it was used by Sandia Labs for atomic bomb fit tests and weapons load training. B-36H (51-5712) was used for the development of the NB-36H nuclear reactor test aircraft. (U.S. Air Force)

Only two of the 82 damaged aircraft were not returned to service. One B-36D (44-92051) was picked up by the tornado and landed in a ravine. The fuselage was broken in half and only one wing remained. The left wing and tail assembly were severed from the aircraft. Parts were used to fix other tornado damaged aircraft. This aircraft was later shipped to Kirtland AFB, NM where it was used by Sandia Labs for atomic bomb fit tests and weapons load training. The other aircraft B-36H (51-5712) was used for the development of the NB-36H nuclear reactor test aircraft.

Broken Arrow
B-36B (44-9275). On 13 February 1950, America's first ever "Broken Arrow"[55] occurred when this B-36B left Eielson AFB, AK enroute to its home base at Carswell AFB, TX. The 16-hour flight included a "simulated combat profile" mission. The route would take it non-stop via Washington State and Montana. Here the B-36 would climb to 40,000 feet over southern California and then do a simulated bomb run over San Francisco. It would then continue its non-stop flight to Fort Worth.

Six hours after takeoff the aircraft experienced icing conditions and multiple engine fires. SAC headquarters at Offutt AFB, NE received several distress messages in quick succession. The first at 11:25 pm said the aircraft was experiencing instrumentation problems at 40,000 feet and was descending to 15,000 feet. A second message reported: "One engine on fire. Contemplate ditching in Queen Charlotte Sound between Queen Charlotte Island and Vancouver Island. Keep a careful lookout for flares or wreckage." Soon two engines caught fire, and one had to be feathered. Shortly after, a fire started in No.5, and then No. 3 stopped with a plugged line.

The pilot, Captain Harold L. Barry, flew the crippled aircraft out over the ocean to release the Mk-4 atomic bomb it was carrying. The bomb was an upgrade of the Mk-3 "Fat Man" bomb dropped over Nagasaki, Japan in 1945. It was onboard to allow the crew to experience the flight characteristics of carrying an atomic bomb and test the ability of the crew to arm the weapon inflight. The bomb had a lead core, and no nuclear material was onboard the aircraft. But the bomb was still top secret and needed to be exploded to keep the Soviets from finding it. The crew dropped it as soon as the aircraft was over the water. It was set to explode at 3,000 feet. When it did the crew could see the fireball from their altitude at 8,000 feet.

The final message indicated that the plane was going to ditch in Queen Charlotte Sound. Barry aimed the aircraft southwest, engaged the autopilot and ordered the crew to bail out. Five crewmen were lost and presumed dead although one was rescued a few days later. Although he escaped his fate that day, ironically, on 27 April 1951 Captain Barry was flying as the copilot of a B-36D (49-2658) when it crashed following a mid-air collision with a P-51 Mustang 50 miles northeast of Oklahoma City, OK. Barry and eleven other B-36 crewmen, along with the Mustang pilot, was killed that day.

An investigation found that ice buildup on the carburetors caused the engines to run rich and eventually caused the fires. Crews searched for the aircraft for days before calling off the search. It was believed at that time to be at the bottom of the ocean. A Royal Canadian Air Force flight over British Columbia discovered the wreck three years later. The United States Air Force sent in search crews in 1953 to no avail. They tried again in 1954 using a Canadian guide and better equipment. This time the team discovered the wreckage. They took critical parts from it and blew up the rest. The crash was discovered again on 23 June 1956 by a 22-year-old participating in a field mapping exercise for the Geological Survey of Canada. Otherwise, the crash was kept secret for 40 years and those who knew about the crash were sworn to secrecy.[56]

B-36J (52-2816). The second Broken Arrow occurred on 22 May 1957, when a 42,000 pound, 10-megaton bomb accidently dropped and detonated outside Kirtland AFB, NM. The B-36J was attempting to land at Kirtland when the bomb was accidentally dropped. The Mk-17 device the aircraft carried was the largest weapon ever made up to that point and the first thermonuclear device designed to be air dropped. The aircraft, commanded by Lieutenant Colonel Richard Meyer, was enroute from Biggs AFB, TX carrying the bomb to Kirtland, ironically, to be permanently disarmed.

Standard operating procedure on all such flights called for the manual removal of the locking pin designed to prevent accidental inflight release of bombs to allow emergency jettisoning of weapons, if necessary, during takeoffs and landings. The awkward procedure required a crew member, usually the navigator, to climb into the bomb bay and lean over the body of the bomb at the start and end of each flight to set and later remove the large pin. On that day, 1st Lieutenant Bob Carp was assigned the onerous task.[57]

Carp entered the bomb bay when the aircraft reached 1,700 feet on its descent. He carefully hung over the huge 25-foot-long bomb literally hanging by his toes to reach the pin. Carp momentarily lost his balance and reached up quickly for a handhold to keep from falling. He instinctively grabbed a lever that immediately let the bomb loose. It went crashing through the closed bomb bay doors and left a huge hole in the bottom of the aircraft. Somehow, Carp held on and remained in the bomb bay as the bomb hurdled toward the earth and the plane shot upward 1,500 feet.

The bomb hit the ground at full speed since there was no time for the retarding parachute to open. It detonated with an earth-shattering explosion. It produced a 12-foot-deep crater, 25 feet in diameter. Luckily, it fell on uninhabited land on a barren mesa. There was no nuclear explosion either. Only the 300 pounds of conventional explosives designed to trigger the nuclear chain reaction exploded. No radioactivity was released.

Displays

The first B-36 to be retired arrived at the MASDC on Davis-Monthan AFB, AZ in February 1956. Since the Air Force had no immediate or planned future need for these aircraft, they were all scrapped over the next three years. By late 1958 only 22 B-36 aircraft were left on active duty. The last B-36 was retired on 12 February 1959 when a B-36J (52-2827) of the 95th BMW was flown from Biggs AFB, TX to Fort Worth for static display. Departure of this B-36 made SAC an all-jet bomber force. All remaining B-36 aircraft, not previously destroyed or identified for display, were scrapped by April 1959.[1] Only four aircraft remain, and all are on static display. There are no flyable examples of the B-36.

B-36J (52-2827)

This B-36J is on display at the Pima Air and Space Museum in Tucson, AZ. It was retired on 12 February 1959 and flew the last operational B-36 flight (Operation SAYONARA) from Biggs AFB, TX to Fort Worth for display at the Ammon Carter Field airport, later renamed the Greater Southwest (GSW) airport, under the Air

Convair B-36J (52-2827) Peacemaker on display at the Pima Air & Space Museum in Tucson, Arizona. (Scott Youmans, courtesy of Pima Air & Space Museum)

Force Museum loan program. This aircraft was not only the last B-36 to fly, but it was also the last one built (MSN 383). It rolled off the assembly line on 1 July 1954 and was delivered to the 92nd BMW at Fairchild AFB, WA on 14 August 1954. It was transferred to the 95th BMW at Biggs AFB, TX on 1 August 1956 where it remained until its retirement. It logged a total of 1,414.8 hours flying time during its service life.

The aircraft was displayed at the GSW airport for several years. It was moved to the Southwest Aero Museum when the GSW airport closed in 1974. A group called the Peacemaker Foundation, headed by a Convair aeronautical engineer named Sam Ball, received permission from the city to restore the aircraft. The group planned to use it as a flying museum and offer flights from Fort Worth's Meacham Field airport. They managed to get the engines running when the Air Force, alarmed that the aircraft might become airworthy, halted the program. The group negotiated with the Air Force for two years. Eventually, the project was abandoned. The museum was subsequently closed, and another group put the aircraft on display along with several other types near Air Force Plant 4 (then operated by General Dynamics) in Fort Worth. Eventually, the aircraft began to fall into decay and neglect.

In 1992, the Aviation Heritage Association was formed to save the aircraft. With the full support of the city, the aircraft was moved into a hangar at the plant now operated by Lockheed Martin. The aircraft was disassembled and restored section by section to its original condition. It was fully painted including the addition of the SAC patch and banner. The area around the canopy bubble was painted white as was the standard for the 95th BMW, Biggs AFB, TX and 6th BMW, Walker AFB, NM. This was intended to reduce the heat in the cockpit caused by the intense sun in those locations.

After more than 12 years and 44,000 man-hours the aircraft restoration was nearly completed, and the aircraft would require about three months to reassemble when a suitable museum location was found. The Heritage Association and the city of Fort Worth planned to create a museum at Alliance Airport on the north side of Fort Worth. However, they were never able to raise the funds needed. Meanwhile, Lockheed Martin notified the foundation that they needed their hangar space for F-16 and F-35 production. The aircraft sections were wrapped in protective plastic coverings and placed in outside storage on the back lot of the plant. Finally, in 2005 the aircraft was trucked to the Pima Air and Space Museum in Tucson where restoration was completed, and the aircraft was permanently placed on display in 2009.

B-36J (52-2220)

This B-36J is on display at the National Museum of the United States Air Force at Wright-Patterson AFB, OH. It made the last flight of a B-36 (non-operational) from Davis-Monthan AFB, AZ to Wright-Patterson for display at the Air Force Museum[2] on 30 April 1959. It rolled off the assembly line on 27 October 1953 and was delivered to the 11th BMW at Carswell AFB, TX on 26 January 1954. It was transferred to the 42nd BMW at Loring AFB, ME

Convair B-36J (52-2220) Peacemaker on display at the National Museum of the United States Air Force at Wright-Patterson AFB, Ohio. (National Museum of the U.S. Air Force)

on 10 December 1954, underwent modification in Fort Worth from March to May 1956, and was finally assigned to the 95th BMW at Biggs AFB, TX on 29 May 1956 where it remained until its retirement. This aircraft retired on 3 February 1959, when it flew from Biggs to the MASDC on Davis-Monthan. It was scheduled for reclamation on 5 February 1959 but was saved from that fate when it was assigned to the museum.

The aircraft features white paint on the bottom that became standard in the mid-1950s and was intended to reduce the effect of heat from thermonuclear explosions. The actual specification called for the paint to be "feathered" resulting in a slightly wavy line rather than the more typical straight line shown here. This aircraft replaced RB-36E (42-13571), which was the original YB-36, and the museum's first B-36 display received on 18 February 1957. The RB-36E was displayed outside at the old museum facility until 1972 when a new indoor facility was completed. By then it had severely decayed, and the museum decided to replace it with the B-36J. The RB-36E was then sold for scrap to the late Walter Soplata of Newbury, OH. Its last known location was on Walter's farm where it was in several pieces among parts of a P-63 and XP-82, and a complete F-82. The RB-36E bomb bay contained a complete P-47N still in its crate.

B-36J (52-2217)

This B-36J is on display at the Strategic Air Command & Aerospace Museum in Ashland, NE. The aircraft rolled off the assembly line on 28 September 1953 and was delivered to the 7th BMW at Carswell AFB, TX on 22 December 1953. It was transferred to the 42nd BMW at Loring AFB, ME on 30 September 1954, underwent modification in Fort Worth from May to August 1956, and was finally assigned to the 95th BMW at Biggs AFB, TX on 10 August 1956 where it remained until its retirement. This aircraft retired on 10 February 1959, when it flew from Biggs to the MASDC on Davis-Monthan AFB, AZ. It was scheduled for reclamation on 12 February 1959 but was spared when it was assigned to the Strategic Air Command Museum[3] in Bellevue, NE just outside Offutt AFB. It flew to the museum on 22 April 1959 where it was placed on display outside the museum and by the 1990s was in a serious state of deterioration. Nonetheless, the aircraft was moved in 1998 to the museum's new location in Ashland, NE where it was fully restored and is now displayed indoors.

Convair B-36J (52-2217) Peacemaker on display at the Strategic Air Command & Aerospace Museum in Ashland, Nebraska. (Strategic Air Command & Aerospace Museum)

RB-36H (51-13730)

This RB-36H is on display at the Castle Air Museum in Atwater, CA. The aircraft rolled off the assembly line on 13 August 1952 and was delivered to the 28th SRW at Rapid City AFB, SD on 25 September 1952 where it remained until its retirement. This aircraft retired in March 1957 when it flew from Rapid City AFB, by then called Ellsworth AFB, to Chanute AFB, IL where it was placed on display. It was erroneously painted as tail number 44-92065, a B-36B later converted to B-36D. When Chanute closed in 1993 the aircraft was disassembled and trans-

Convair RB-36H (51-13730) Peacemaker on display at the Castle Air Museum in Atwater, California. (Nehrams2020)

ported by rail to the museum in Atwater. By 1994 the aircraft was reassembled and restored for display where it remains. Its markings reflect the 28th SRW.

Appendix A: Specifications

B-36A Peacemaker

The B-36A was fielded as a 6-engine aircraft using the "pusher" propeller configuration. (U.S. Air Force)

Aircraft Specifications					
Type:	B-36	Series:	B-36A	Name:	Peacemaker
Wingspan:	230.0 feet	Length:	162.1 feet	Height:	46.8 feet
Empty Weight:	135,020 pounds	Combat Weight:	212,800 pounds	Max TO Weight:	310,380 pounds
Combat Radius:	3,880 miles	Combat Ceiling:	35,800 feet	Service Ceiling:	39,100 feet
Cruise Speed:	218 mph	Max Speed:	345 mph	Max Payload:	72,000 pounds
OEM:	Convair	Produced:	22	SAC Inventory:	20
First Flight:	28-Aug-47	First Delivery:	30-Aug-47	Phase Out:	Jul-51
Missions:	Intercontinental Strategic Bombardment					
Tail Number(s):	44-92004 to 44-92025					
Propulsion:	Six Pratt & Whitney R-4360-25 Wasp Major radials of 3,000 hp each.					
Accommodations:	Total 15. Fwd: Pilot, Copilot, Navigator, Radar-Bombardier, Engineer, (2) Radio Operators, Nose Gun, Lt Gun, and Rt Gun. Aft: Upper Lt/Rt Gun, Lower Lt/Rt Gun, and Tail Gun (gunners acted primarily as observers since no guns were installed).					
Payload:	Up to 72,000 pounds of conventional bombs (not atomic bomb capable) in four bomb bays plus sixteen 20mm cannons (no defensive armaments installed).					
Comments:	Combat radius was 3,880 miles with 10,000-pound bombload, and 2,100 miles with 72,000-pound bombload. Max speed was 345 mph at 31,600 feet. Combat ceiling was 35,800 feet with 10,000-pound bombload. One B-36A (44-92004) was tested to destruction (designated YB-36A). Another B-36A (44-92005) was converted to EB-36A. SAC took possession of 20 B-36A aircraft. B-36A aircraft were unarmed and used for crew training only. All B-36A aircraft (including 44-92005) and the YB-36 were converted to RB-36E (22 total).					

B-36B Peacemaker

The B-36B was built with 6 engines but later converted to B-36D with 4 jets added. (U.S. Air Force)

Aircraft Specifications					
Type:	B-36	Series:	B-36B	Name:	Peacemaker
Wingspan:	230.0 feet	Length:	162.1 feet	Height:	46.8 feet
Empty Weight:	140,640 pounds	Combat Weight:	227,700 pounds	Max TO Weight:	328,000 pounds
Combat Radius:	4,307 miles	Combat Ceiling:	38,800 feet	Service Ceiling:	42,500 feet
Cruise Speed:	203 mph	Max Speed:	381 mph	Max Payload:	86,000 pounds
OEM:	Convair	Produced:	62	SAC Inventory:	54
First Flight:	8-Jul-48	First Delivery:	30-Nov-48	Phase Out:	Feb-52
Missions:	Intercontinental Strategic Bombardment				
Tail Number(s):	44-92026 to 44-92087 (44-92088 to 44-92094 delivered as RB-36D; 44-92095 to 44-92098 as B-36D)				
Propulsion:	Six Pratt & Whitney R-4360-41 Wasp Major radials of 3,500 hp each.				
Accommodations:	Total 15. Fwd: Pilot, Copilot, Navigator, Radar-Bombardier, Engineer, Radio Operator, Nose Gun, Lt Gun, and Rt Gun. Aft: Upper Lt/Rt Gun, Lower Lt/Rt Gun, Tail Gun, and Auxiliary Crew Member.				
Payload:	Up to 86,000 pounds of bombs (including two 43,000-pound T-12 Cloudmaker conventional or two 42,000-pound Mk-17 atomic bombs) plus sixteen 20mm cannons.				
Comments:	Some aircraft limited to 278,000 pounds max TO weight due to landing gear restrictions. Combat radius was 4,307 miles with 10,000-pound bombload. Max speed was 381 mph at 34,500 feet. Combat ceiling was 38,800 feet with 10,000-pound bombload. Total of 47 aircraft (44-92045 to 44-92087) later modified to carry the Mk-III/4 atomic bomb. Production of 73 aircraft ordered with 62 delivered as B-36B (Convair converted and delivered 4 as B-36D and 7 as RB-36D). Production included 34 aircraft originally planned as B-36C. A total of 59 B-36B aircraft were subsequently modified with four J47 jet engines and converted to B-36D (3 crashed as B-36B before conversions took place).				

B-36D Peacemaker

The B-36D was essentially a B-36B modified with 4 jet engines and other performance improvements. (U.S. Air Force)

Aircraft Specifications					
Type:	B-36	Series:	B-36D	Name:	Peacemaker
Wingspan:	230.0 feet	Length:	162.1 feet	Height:	46.8 feet
Empty Weight:	160,974 pounds	Combat Weight:	248,410 pounds	Max TO Weight:	357,500 pounds
Combat Radius:	3,529 miles	Combat Ceiling:	41,850 feet	Service Ceiling:	44,300 feet
Cruise Speed:	221 mph	Max Speed:	406 mph	Max Payload:	86,000 pounds
OEM:	Convair	Produced:	85	SAC Inventory:	82
First Flight:	11-Jul-49	First Delivery:	22-Aug-50	Phase Out:	Mar-57
Missions:	Intercontinental Strategic Bombardment of Ground and Naval Material Objectives				
Tail Number(s):	44-92026 to 44-92034, 44-92036 to 44-92074, 44-92076 to 44-92078, 44-92080 to 44-92087, 44-92094 to 44-92098, 49-2647 to 49-2668				
Propulsion:	Six Pratt & Whitney R-4360-41 radials of 3,500 hp each and four General Electric J47-GE-19 turbojets of 5,010 pounds thrust each.				
Accommodations:	Total 15. Fwd: Aircraft Commander, Pilot, Copilot (Lt Gun), Navigator, Radar-Bombardier, (2) Engineers, (2) Radio/ECM Op (Rt Gun), Observer (Nose Gun). Aft: Upper Lt/Rt Gun, Lower Lt/Rt Gun, Tail Gun.				
Payload:	Up to 86,000 pounds of bombs (including two 43,000-pound T-12 Cloudmaker conventional or two 42,000-pound Mk-17 atomic bombs) plus sixteen 20mm cannons.				
Comments:	Max TO weight increased to 370,000 pounds for aircraft modified with improved landing gear. Combat radius was 3,529 miles with 10,000-pound bombload. Max speed was 406 mph at 36,200 feet. Combat ceiling was 41,850 feet with 10,000-pound bombload. Total 85 aircraft produced (22 new production plus 4 B-36B converted in production, and 59 converted from operational B-36B). First flight with J35 engines on 26-Mar-49. First new-build B-36D flew on 11-Jul-49. B-47 engine pod design used with Boeing approval. First conversion completed 5-Oct-50 and first new-build delivered 22-Aug-50.				

RB-36D Peacemaker

The RB-36D was easily recognizable by the bright aluminum skin covering the camera compartment. (U.S. Air Force)

Aircraft Specifications					
Type:	B-36	Series:	RB-36D	Name:	Peacemaker
Wingspan:	230.0 feet	Length:	162.1 feet	Height:	46.8 feet
Empty Weight:	165,029 pounds	Combat Weight:	258,675 pounds	Max TO Weight:	370,000 pounds
Combat Radius:	3,494 miles	Combat Ceiling:	40,000 feet	Service Ceiling:	43,400 feet
Cruise Speed:	219 mph	Max Speed:	401 mph	Max Payload:	86,000 pounds
OEM:	Convair	Produced:	24	SAC Inventory:	23
First Flight:	18-Dec-49	First Delivery:	3-Jun-50	Phase Out:	Sep-56
Missions:	Strategic Reconnaissance, Mapping, Charting, Bomb Damage Assessment				
Tail Number(s):	44-92088 to 44-92094, 49-2686 to 49-2702				
Propulsion:	Six Pratt & Whitney R-4360-41 radials of 3,500 hp each and four General Electric J47-GE-19 turbojets of 5,010 pounds thrust each.				
Accommodations:	Total 22. Fwd: Aircraft Commander, Pilot, Copilot (Lt Gun), Navigator, Photo-Navigator, Radar-Observer, (2) Engineers, (2) Radio Op (Rt Gun), (4) ECM Op, (2) Photo Tech, Observer (Nose Gun). Aft: Upper Lt/Rt Gun, Lower Lt/Rt Gun, Tail Gun.				
Payload:	80 T-86 photoflash bombs, cameras (4 K-17C, 3 K-22A, 7 K-38) and alternates (1 K-17C, 1 K-22A, 1 K-37, 7 K-38, 5 K-40, and 1 T-11) plus sixteen 20mm cannons.				
Comments:	Combat radius was 3,494 miles with 10,000-pound payload. Max speed was 401 mph at 36,500 feet. Combat ceiling 40,000 feet with 10,000-pound payload. One aircraft (44-92088) spent its entire service as a test aircraft. It was redesignated ERB-36D in 1954 and fitted with the Boston Camera. Additional fuel tank carried 3,000 gallons and increased endurance to 50 hours. Unofficial operational ceiling was 50,000 feet (58,000 for Featherweight version). Distinguished by the bright aluminum skin covering the camera compartment and by several radomes under the aft fuselage, varying in number and placement.				

RB-36E Peacemaker

The RB-36E was a B-36A quickly converted for reconnaissance missions to meet LeMay's urgent need. (U.S. Air Force)

	Aircraft Specifications				
Type:	B-36	Series:	RB-36E	Name:	Peacemaker
Wingspan:	230.0 feet	Length:	162.1 feet	Height:	46.7 feet
Empty Weight:	164,238 pounds	Combat Weight:	258,200 pounds	Max TO Weight:	370,000 pounds
Combat Radius:	3,520 miles	Combat Ceiling:	40,000 feet	Service Ceiling:	43,400 feet
Cruise Speed:	219 mph	Max Speed:	401 mph	Max Payload:	86,000 pounds
OEM:	Convair	Produced:	22	SAC Inventory:	22
First Flight:	7-Jul-50	First Delivery:	31-Jul-50	Phase Out:	Feb-57
Missions:	Strategic Reconnaissance, Mapping, Charting, Bomb Damage Assessment				
Tail Number(s):	42-13571, 44-92005 to 44-92025				
Propulsion:	Six Pratt & Whitney R-4360-41 radials of 3,500 hp each and four General Electric J47-GE-19 turbojets of 5,010 pounds thrust each.				
Accommodations:	Total 22. Fwd: Aircraft Commander, Pilot, Copilot (Lt Gun), Navigator, Photo-Navigator, Radar-Observer, (2) Engineers, (2) Radio Op (Rt Gun), (4) ECM Op, (2) Photo Tech, Observer (Nose Gun). Aft: Upper Lt/Rt Gun, Lower Lt/Rt Gun, Tail Gun.				
Payload:	80 T-86 photoflash bombs, cameras (4 K-17C, 3 K-22A, 7 K-38) and alternates (1 K-17C, 1 K-22A, 1 K-37, 7 K-38, 5 K-40, and 1 T-11) plus sixteen 20mm cannons.				
Comments:	Combat radius was 3,520 miles with 10,000-pound payload. Max speed was 401 mph at 36,500 feet. Combat ceiling 40,000 feet with 10,000-pound payload. Total 22 aircraft produced (1 converted from YB-36, and 21 converted from B-36A). The conversion included an upgrade of the engines from the R-4360-25 to the more powerful R-4360-41 engines used on the B-36B. They were also fitted with four J47 jet engines as used on the RB-36D. Bomb bay configuration and crew complement was the same as the RB-36D. Unofficial operational ceiling was 50,000 feet (58,000 for Featherweight version).				

B-36F Peacemaker

The B-36F was improved B-36D with more powerful engines and max speed of 418 mph. (National Museum of the USAF)

Aircraft Specifications					
Type:	B-36	Series:	B-36F	Name:	Peacemaker
Wingspan:	230.0 feet	Length:	162.1 feet	Height:	46.8 feet
Empty Weight:	167,646 pounds	Combat Weight:	254,300 pounds	Max TO Weight:	370,000 pounds
Combat Radius:	3,232 miles	Combat Ceiling:	40,900 feet	Service Ceiling:	44,000 feet
Cruise Speed:	235 mph	Max Speed:	418 mph	Max Payload:	86,000 pounds
OEM:	Convair	Produced:	34	SAC Inventory:	34
First Flight:	18-Nov-50	First Delivery:	31-Mar-51	Phase Out:	Aug-57
Missions:	Intercontinental Strategic Bombardment of Ground and Naval Material Objectives				
Tail Number(s):	49-2669 to 49-2675, 49-2677 to 49-2683, 49-2685, 50-1064 to 50-1082				
Propulsion:	Six Pratt & Whitney R-4360-53 radials of 3,800 hp each and four General Electric J47-GE-19 turbojets of 5,010 pounds thrust each.				
Accommodations:	Total 15. Fwd: Aircraft Commander, Pilot, Copilot (Lt Gun), Navigator, Radar-Bombardier, (2) Engineers, (2) Radio/ECM Op (Rt Gun), Observer (Nose Gun). Aft: Upper Lt/Rt Gun, Lower Lt/Rt Gun, Tail Gun.				
Payload:	Up to 86,000 pounds of bombs (including two 43,000-pound T-12 Cloudmaker conventional or two 42,000-pound Mk-17 atomic bombs) plus sixteen 20mm cannons.				
Comments:	Improved version of B-36D. Combat radius was 3,232 miles with 10,000-pound bombload. Max speed was 418 mph at 37,100 feet. Combat ceiling 44,000 feet with 10,000-pound bombload. Avionics improvements included K-3A radar bombing navigation system, AN/APG-32A defensive fire control system, and chaff dispensing system. Two aircraft (49-2676 and 49-2684) removed from B-36F production and delivered as XB-36G (later YB-60) all jet bomber with 8 jets mounted on 4 pods and swept wing like the B-52. First aircraft delivered to SAC (49-2671) delayed until 18 August 1951 due to accelerated test program.				

RB-36F Peacemaker

The RB-36F (49-2707) shown became the FICON prototype and was redesignated JRB-36F. (U.S. Air Force)

Aircraft Specifications					
Type:	B-36	Series:	RB-36F	Name:	Peacemaker
Wingspan:	230.0 feet	Length:	162.1 feet	Height:	46.8 feet
Empty Weight:	170,889 pounds	Combat Weight:	262,800 pounds	Max TO Weight:	370,000 pounds
Combat Radius:	3,098 miles	Combat Ceiling:	40,100 feet	Service Ceiling:	43,100 feet
Cruise Speed:	228 mph	Max Speed:	409 mph	Max Payload:	86,000 pounds
OEM:	Convair	Produced:	24	SAC Inventory:	23
First Flight:	30-Apr-51	First Delivery:	8-May-51	Phase Out:	Oct-58
Missions:	Strategic Reconnaissance, Mapping, Charting, Bomb Damage Assessment				
Tail Number(s):	49-2703 to 49-2721, 50-1098 to 50-1102				
Propulsion:	Six Pratt & Whitney R-4360-53 radials of 3,800 hp each and four General Electric J47-GE-19 turbojets of 5,010 pounds thrust each.				
Accommodations:	Total 22. Fwd: Aircraft Commander, Pilot, Copilot (Lt Gun), Navigator, Photo-Navigator, Radar-Observer, (2) Engineers, (2) Radio Op (Rt Gun), (4) ECM Op, (2) Photo Tech, Observer (Nose Gun). Aft: Upper Lt/Rt Gun, Lower Lt/Rt Gun, Tail Gun.				
Payload:	80 T-86 photoflash bombs, cameras (4 K-17C, 3 K-22A, 7 K-38) and alternates (1 K-17C, 1 K-22A, 1 K-37, 7 K-38, 5 K-40, and 1 T-11) plus sixteen 20mm cannons.				
Comments:	Combat radius was 3,098 miles with 10,000-pound payload. Max speed was 409 mph at 36,100 feet. Combat ceiling 40,100 feet with 10,000-pound payload. The first aircraft delivered (49-2707) was assigned to AMC on 8 May 1951 for testing the FICON concept and was redesignated JRB-36F. The first aircraft assigned to SAC was delivered to the 5th SRW at Travis AFB, CA on 28 May 1951. All aircraft were initially assigned to the 5th SRW, and later transferred to the 99th SRW at Fairchild AFB, WA and finally to the 72nd SRW at Ramey AFB, Puerto Rico. Unofficial operational ceiling was 50,000 feet (58,000 for Featherweight version).				

B-36H Peacemaker

The B-36H was an improved B-36F with 2nd FE station and improved tail turret fire control system. (U.S. Air Force)

Aircraft Specifications

Type:	B-36	Series:	B-36H	Name:	Peacemaker
Wingspan:	230.0 feet	Length:	162.1 feet	Height:	46.8 feet
Empty Weight:	168,487 pounds	Combat Weight:	253,900 pounds	Max TO Weight:	370,000 pounds
Combat Radius:	3,113 miles	Combat Ceiling:	40,800 feet	Service Ceiling:	44,000 feet
Cruise Speed:	234 mph	Max Speed:	416 mph	Max Payload:	86,000 pounds
OEM:	Convair	Produced:	83	SAC Inventory:	79
First Flight:	5-Apr-52	First Delivery:	31-Dec-51	Phase Out:	Jun-58
Missions:	Intercontinental Strategic Bombardment of Ground and Naval Material Objectives				
Tail Number(s):	50-1083 to 50-1087, 51-5699 to 51-5742, 52-1343 to 52-1366				
Propulsion:	Six Pratt & Whitney R-4360-53 radials of 3,800 hp each and four General Electric J47-GE-19 turbojets of 5,010 pounds thrust each.				
Accommodations:	Total 15. Fwd: Aircraft Commander, Pilot, Copilot (Lt Gun), Navigator, Radar-Bombardier, (2) Engineers, (2) Radio/ECM Op (Rt Gun), Observer (Nose Gun). Aft: Upper Lt/Rt Gun, Lower Lt/Rt Gun, Tail Gun.				
Payload:	Up to 86,000 pounds of bombs (including two 43,000-pound T-12 Cloudmaker conventional or two 42,000-pound Mk-17 atomic bombs) plus sixteen 20mm cannons.				
Comments:	Improved version of B-36F. Combat radius was 3,113 miles with 10,000-pound bombload. Max speed was 416 mph at 36,700 feet. Combat ceiling 40,800 feet with 10,000-pound bombload. Improvements included a second FE station, radar relocated to pressurized compartment, and AN/APG-41A defensive fire control system (tail turret). One B-36H (51-5712) damaged in the Carswell AFB tornado on 1 September 1952 was converted to XB-36H, and then NB-36H, and used for nuclear reactor testing. Several aircraft were used for experiments including an aerial tanker and GAM Director, and atomic test support.				

RB-36H Peacemaker

The RB-36H had similar performance to the RB-36F and incorporated most B-36H design improvements. (U.S. Air Force)

Aircraft Specifications					
Type:	B-36	Series:	RB-36H	Name:	Peacemaker
Wingspan:	230.0 feet	Length:	162.1 feet	Height:	46.8 feet
Empty Weight:	171,942 pounds	Combat Weight:	263,300 pounds	Max TO Weight:	370,000 pounds
Combat Radius:	2,936 miles	Combat Ceiling:	39,900 feet	Service Ceiling:	42,700 feet
Cruise Speed:	229 mph	Max Speed:	408 mph	Max Payload:	86,000 pounds
OEM:	Convair	Produced:	73	SAC Inventory:	73
First Flight:	Jan-52	First Delivery:	7-Feb-52	Phase Out:	Nov-58
Missions:	Strategic Reconnaissance, Mapping, Charting, Bomb Damage Assessment				
Tail Number(s):	50-1103 to 50-1110, 51-13717 to 51-13741, 51-5743 to 51-5756, 52-1367 to 52-1392				
Propulsion:	Six Pratt & Whitney R-4360-53 radials of 3,800 hp each and four General Electric J47-GE-19 turbojets of 5,010 pounds thrust each.				
Accommodations:	Total 22. Fwd: Aircraft Commander, Pilot, Copilot (Lt Gun), Navigator, Photo-Navigator, Radar-Observer, (2) Engineers, (2) Radio Op (Rt Gun), (4) ECM Op, (2) Photo Tech, Observer (Nose Gun). Aft: Upper Lt/Rt Gun, Lower Lt/Rt Gun, Tail Gun.				
Payload:	80 T-86 photoflash bombs, cameras (4 K-17C, 3 K-22A, 7 K-38) and alternates (1 K-17C, 1 K-22A, 1 K-37, 7 K-38, 5 K-40, and 1 T-11) plus sixteen 20mm cannons.				
Comments:	Combat radius was 2,936 miles with 10,000-pound payload. Max speed was 408 mph at 36,750 feet. Combat ceiling 39,900 feet with 10,000-pound payload. This aircraft had similar performance to the RB-36F and incorporated most of the design improvements of the standard B-36H. Although aircraft deliveries did not begin until 1952, the original contract was signed in 1950 resulting in the aircraft carrying 1950 series tail numbers. Unofficial operational ceiling was 50,000 feet (58,000 for Featherweight version).				

B-36J Peacemaker

The B-36J had additional fuel cells and strengthened landing gear that increased GW by 50,000 pounds. (U.S. Air Force)

Aircraft Specifications					
Type:	B-36	Series:	B-36J	Name:	Peacemaker
Wingspan:	230.0 feet	Length:	162.1 feet	Height:	46.8 feet
Empty Weight:	171,035 pounds	Combat Weight:	266,100 pounds	Max TO Weight:	410,000 pounds
Combat Radius:	3,403 miles	Combat Ceiling:	39,900 feet	Service Ceiling:	43,000 feet
Cruise Speed:	228 mph	Max Speed:	411 mph	Max Payload:	86,000 pounds
OEM:	Convair	Produced:	33	SAC Inventory:	33
First Flight:	Jul-53	First Delivery:	9-Sep-53	Phase Out:	12-Feb-59
Missions:	Intercontinental Strategic Bombardment of Ground and Naval Material Objectives				
Tail Number(s):	52-2210 to 52-2226, 52-2812 to 52-2827				
Propulsion:	Six Pratt & Whitney R-4360-53 radials of 3,800 hp each and four General Electric J47-GE-19 turbojets of 5,010 pounds thrust each.				
Accommodations:	Total 15. Fwd: Aircraft Commander, Pilot, Copilot (Lt Gun), Navigator, Radar-Bombardier, (2) Engineers, (2) Radio/ECM Op (Rt Gun), Observer (Nose Gun). Aft: Upper Lt/Rt Gun, Lower Lt/Rt Gun, Tail Gun.				
Payload:	Up to 86,000 pounds of bombs (including two 43,000-pound T-12 Cloudmaker conventional or two 42,000-pound Mk-17 atomic bombs) plus sixteen 20mm cannons.				
Comments:	Improved version of B-36H. Combat radius was 3,403 miles with 10,000-pound bombload. Max speed was 411 mph at 36,400 feet. Combat ceiling 43,000 feet with 10,000-pound bombload. Improvements included two additional wing tanks to allow 2,770 gallons of additional fuel, giving a total capacity of 36,396 gallons. The additional fuel load increased the combat radius by approximately 460 miles. It also included stronger landing gear that increased gross weight to 410,000 pounds.				

Appendix B: Colors and Markings

All B-36A and B-models were originally delivered with bare-metal finish and minimal markings. They included a "Buzz Number" on the side of the fuselage (standard B-36 designator "BM" and the last three digits of the tail number), a small tail number centered on the tail, standard national emblem on both sides of the rear fuselage as well as the top of the left wing and bottom of the right wing. Some A-models also had the Buzz Number on bottom of the left wing. (American Aviation Historical Society)

GEM-modified B-36B aircraft and a few A-models were painted bright red on the wingtips and tail so they could be located during Arctic white-out conditions or if they were forced down in remote snow-covered terrain. B-models also added "USAF" markings on the top of the right wing and bottom of the left wing (which eliminated the buzz number from bottom of the left wing). (U.S. Air Force)

The red paint was eliminated for B-36D and later aircraft. The Buzz Numbers were removed and replaced with a smaller "UNITED STATES AIR FORCE" in 9-inch letters. The MSN was painted on each side of the fuselage just aft of the nose. The national emblems on the wings were moved further outboard of the new jet engines. (University of North Texas Libraries)

The Buzz Number (without the BM designator) was retained on some aircraft and moved forward on the fuselage or painted just below the "UNITED STATES AIR FORCE" marking. (University of North Texas Library)

Tail codes like those used in WWII were added. Codes included 5th BMW Circle X, 6th BMW Triangle R, 7th BMW Triangle J, 9th SRW Circle X, 11th BMW Triangle U, 28th SRW Triangle X, 72nd BMW Square F, 92nd BMW Circle W, 99th SRW Circle I. These codes were phased out by the end of 1952 before the 42nd and 95th BMW were activated. Note the Buzz Number painted below the "UNITED STATES AIR FORCE" marking and the unit patch added to the forward fuselage. (U.S. Air Force)

The SAC Shield with the blue "Milky Way" banner became standard on all SAC aircraft including the B-36. The right side initially included just the banner but eventually the unit patch was applied over it like the SAC Shield was on the left side. Notice the underside of the aircraft is also painted with white heat-resistant paint. (University of North Texas Libraries)

The undersides of all B-36 aircraft were eventually painted with white heat-resistant paint because of the extreme temperatures experienced during Operation IVY testing. Also, in late 1955 the "UNITED STATES AIR FORCE" marking was determined to be too small and was replaced with a larger "U.S. AIR FORCE" marking using 36-inch letters. Some aircraft retained the Buzz Numbers and moved them just forward of the "U.S. AIR FORCE" marking or placed them on the forward fuselage in 12-inch letters. The "USAF" marking on the bottom of the left wing was deleted. (American Aviation Historical Society)

B-36A (44-92006) was the first and one of three B-36 aircraft to carry the name "City of Fort Worth". (University of North Texas Libraries)

Appendix C: Tail Numbers

Tail #	CN	Lineage	Status	Last Unit	Date	Comments
42-13570		XB-36	Retired	Convair FW	6-Nov-52	Original B-36 prototype. First flew on 8-Aug-46 and remained at Convair-FW for testing. Delivered to Air Force in Jun-48. Used for various tests including tracked landing gear and NB-36H mock-up. Returned to Convair-FW in Nov-52 for storage and salvage. Donated to Carswell AFB, TX for fire fighter training and later scrapped in May-57.
42-13571		YB-36 RB-36E	Display	72nd BMW	18-Feb-57	Displayed at Air Force Museum, Wright-Patterson AFB, OH. Replaced with B-36J 52-2220 on 30-Apr-59 and sold to Walter Soplata for scrap in 1971.
43-52436		XC-99	Storage	AMARG		Retired and placed on display at Kelly AFB, TX. Deteriorated and donated to AF Museum in 1993. Disassembled in 2004 and transferred to AF Museum in Dayton, OH. Moved to AMARG in 2012 and placed in storage awaiting funding for restoration.
44-92004	1	YB-36A B-36A	Destroyed	Wright ADC	30-Aug-47	Delivered 30-Aug-47. One-time flight to Wright Air Development Center (ADC) Wright Field, OH. Tested to destruction. No operational Service.
44-92005	2	YB-36A B-36A EB-36A B-36A RB-36E RB-36E-III	Retired	72nd BMW	4-Feb-57	Delivered 18-Jun-48 to Air Force Proving Ground, Eglin AFB, FL. Initially designated YB-36A then B-36A before delivery. First production representative B-36A. To reclamation 18-Apr-57.
44-92006	3	YB-36A B-36A RB-36E RB-36E-III	Retired	72nd BMW	13-Feb-57	Initially designated YB-36A then B-36A before delivery. To reclamation 8-Jul-57.
44-92007	4	YB-36A B-36A RB-36E RB-36E-III	Retired	72nd BMW	5-Feb-57	Initially designated YB-36A then B-36A before delivery. Made the first long-range flight with a completely Air Force crew on 22-Aug-48. To reclamation 18-Apr-57.
44-92008	5	YB-36A B-36A RB-36E RB-36E-III	Retired	72nd BMW	19-Sep-56	Initially designated YB-36A then B-36A before delivery. To reclamation 18-Feb-57.
44-92009	6	YB-36A B-36A RB-36E RB-36E-III	Retired	72nd BMW	6-Feb-57	Initially designated YB-36A then B-36A before delivery. To reclamation 18-Apr-57.
44-92010	7	YB-36A B-36A RB-36E RB-36E-III	Retired	72nd BMW	18-Feb-57	Initially designated YB-36A then B-36A before delivery. To reclamation 8-Jul-57.
44-92011	8	YB-36A B-36A	Retired	72nd BMW	12-Feb-57	Initially designated YB-36A then B-36A before delivery. To reclamation 29-Apr-57.

Tail #	CN	Lineage	Status	Last Unit	Date	Comments
		RB-36E RB-36E-III				
44-92012	9	YB-36A B-36A RB-36E RB-36E-III	Retired	72nd BMW	18-Sep-56	Initially designated YB-36A then B-36A before delivery. To reclamation 11-Dec-56.
44-92013	10	YB-36A B-36A RB-36E RB-36E-II	Retired	72nd BMW	26-Sep-56	Initially designated YB-36A then B-36A before delivery. Made three long-distance record flights in 1948. To reclamation 15-Feb-57.
44-92014	11	YB-36A B-36A RB-36E RB-36E-II	Retired	72nd BMW	24-Sep-56	Initially designated YB-36A then B-36A before delivery. To reclamation 11-Dec-56.
44-92015	12	YB-36A B-36A RB-36E RB-36E-II	Retired	72nd BMW	17-Sep-56	Initially designated YB-36A then B-36A before delivery. To reclamation 11-Dec-56.
44-92016	13	YB-36A B-36A RB-36E RB-36E-II	Retired	72nd BMW	11-Feb-57	Initially designated YB-36A then B-36A before delivery. To reclamation 29-Apr-57.
44-92017	14	B-36A RB-36E RB-36E-II	Retired	72nd BMW	12-Sep-56	To reclamation 15-Feb-57.
44-92018	15	B-36A RB-36E RB-36E-II	Retired	72nd BMW	7-Feb-57	To reclamation 29-Apr-57.
44-92019	16	B-36A RB-36E RB-36E-II	Retired	72nd BMW	31-Jul-56	To reclamation 11-Dec-56.
44-92020	17	B-36A RB-36E RB-36E-II	Retired	72nd BMW	11-Sep-56	To reclamation 11-Dec-56.
44-92021	18	B-36A RB-36E RB-36E-III	Retired	72nd BMW	2-Aug-56	To reclamation 10-Oct-57.
44-92022	19	B-36A RB-36E RB-36E-II	Retired	72nd BMW	12-Aug-56	To reclamation 10-Oct-57.
44-92023	20	B-36A RB-36E RB-36E-II	Retired	72nd BMW	17-Sep-56	To reclamation 11-Dec-56.
44-92024	21	B-36A RB-36E RB-36E-II	Retired	72nd BMW		First RB-36E delivered to SAC (28th SRW) on 31-Jul-50. Retired to Davis-Monthan AFB, AZ and used for fire fighter training. Retirement and reclamation dates unknown.
44-92025	22	B-36A RB-36E RB-36E-II	Retired	72nd BMW	14-Feb-57	To reclamation 8-Jul-57.

Tail #	CN	Lineage	Status	Last Unit	Date	Comments
44-92026	23	YB-36B B-36B B-36D B-36D-III	Retired	95th BMW	4-Jun-56	B-36B prototype. First flew on 8-Jul-48. One of the first four B-36D conversions. Completed in Fort Worth. To reclamation 10-Sep-56.
44-92027	24	B-36B B-36D B-36D-III	Retired	95th BMW	7-Mar-56	To reclamation 10-Sep-56.
44-92028	25	B-36B B-36D B-36D-III	Retired	95th BMW	8-May-56	To reclamation 10-Sep-56.
44-92029	26	B-36B B-36D B-36D-III	Destroyed	95th BMW	8-Feb-55	Crashed 7 miles NE of Biggs AFB, TX on 16-Jul-54. Later returned to service. Damaged beyond repair on landing short of runway at Carswell AFB, TX and written off on 8-Feb-55. Landing gear was sheared off. Total 2 fatalities (all crew).
44-92030	27	B-36B B-36D B-36D-II	Destroyed	42nd BMW	6-Mar-55	Damaged during landing at Walker AFB, NM on 9-Feb-53. Later returned to service. Received structural damage at Loring AFB, ME on 13-Apr-54 when wingtip hit a snowbank on landing. Written off 6-Mar-55. No fatalities, 11 survivors.
44-92031	28	B-36B B-36D B-36D-II	Retired	42nd BMW	28-Feb-56	To reclamation 10-Sep-56.
44-92032	29	B-36B B-36D B-36D-II	Destroyed	92nd BMW	29-Mar-54	Crashed during practice take-off. Aircraft made sudden right turn and hit the ground, narrowly missed several other aircraft, plowed through an unoccupied building, and burst into flames. Total 7 fatalities (all crew), 3 survivors.
44-92033	30	B-36B B-36D B-36D-II	Retired	92nd BMW	19-Nov-56	To reclamation 12-Mar-57.
44-92034	31	B-36B B-36D B-36D-II	Retired	92nd BMW	24-Nov-56	One of the first four B-36D conversions. Completed in Fort Worth. To reclamation 28-Mar-57.
44-92035	32	B-36B	Destroyed	7th BMW	22-Nov-50	Set a long-distance record for a 9,600-mile flight in 43 hours 37 minutes without refueling on 12-Mar-49. Crashed 20 miles south of Carswell AFB, TX on 22-Nov-50. Vibrations from Gunnery practice caused electronic engine control failure. Engines deteriorated on return flight. Crew abandoned the aircraft when it lost airspeed and altitude. Crew could not jettison bombs and fuel due to electrical failure. Total 2 fatalities (all crew), 15 survivors.
44-92036	33	B-36B B-36D B-36D-III	Retired	95th BMW	10-Sep-56	To reclamation 11-Dec-56.

Tail #	CN	Lineage	Status	Last Unit	Date	Comments
44-92037	34	B-36B B-36D B-36D-II	Retired	42nd BMW	8-Feb-56	To reclamation 10-Sep-56.
44-92038	35	B-36B B-36D	Destroyed	Convair SD	12-Jun-52	Damaged on take-off from Davis-Monthan AFB, AZ on 19-Jan-52. Later returned to service. Destroyed by fire on ground at the Convair plant in San Diego and written off on 12-Jun-52. No fatalities (no personnel onboard).
44-92039	36	B-36B B-36D B-36D-III	Retired	95th BMW	9-Jul-56	To reclamation 12-Mar-57.
44-92040	37	B-36B B-36D B-36D-III	Retired	95th BMW	4-Sep-56	To reclamation 11-Dec-56.
44-92041	38	B-36B B-36D B-36D-III	Destroyed	95th BMW	19-Jan-56	Crashed due to mechanical failure 4 miles NW of Fort Worth, TX on 29-Jul-49. Later returned to service. Damaged beyond repair due to hard landing at Biggs AFB, TX and written off on 19-Jan-56. No fatalities.
44-92042	39	B-36B B-36D B-36D-III	Retired	95th BMW	19-May-56	To reclamation 10-Sep-56.
44-92043	40	B-36B B-36D B-36D-II	Retired	42nd BMW	13-Feb-56	To reclamation 10-Sep-56.
44-92044	41	B-36B B-36D B-36D-III	Retired	95th BMW	15-Aug-56	To reclamation 13-Sep-57.
44-92045	42	B-36B B-36D B-36D-III	Retired	95th BMW	14-Aug-56	To reclamation 4-Sep-56.
44-92046	43	B-36B B-36D B-36D-III	Retired	95th BMW	22-May-56	Second B-36B equipped with J47 jet engines and used for high altitude tests. To reclamation 10-Sep-56.
44-92047	44	B-36B B-36D B-36D-II	Retired	42nd BMW	24-Feb-56	To reclamation 10-Sep-56.
44-92048	45	B-36B B-36D B-36D-II	Retired	92nd BMW	4-Dec-56	To reclamation 12-Feb-57.
44-92049	46	B-36B B-36D B-36D-III	Retired	95th BMW	18-Jun-56	To reclamation 10-Sep-56.
44-92050	47	B-36B B-36D	Destroyed	92nd BMW	15-Apr-52	Crashed at end of runway during max weight take-off from Fairchild AFB, WA. Trim incorrectly set and crew tried to correct during take-off roll. Tried to abort take-off by cutting fuel to engines. Total 15 fatalities (all crew), 2 survivors.

Tail #	CN	Lineage	Status	Last Unit	Date	Comments
44-92051	48	B-36B B-36D	Destroyed	11th BMW	1-Sep-52	Damaged by tornado at Carswell AFB, TX. Not returned to service. Parts used to fix other tornado damaged aircraft. Later shipped to Sandia Labs, NM for atomic bomb fit tests and weapons load trainer. No fatalities.
44-92052	49	B-36B B-36D B-36D-III	Retired	95th BMW	7-Aug-56	To reclamation 30-Sep-57.
44-92053	50	B-36B B-36D B-36D-III	Retired	95th BMW	10-Sep-56	One of the first four B-36D conversions. Completed in Fort Worth. To reclamation 11-Dec-56.
44-92054	51	B-36B B-36D JB-36D	Retired	Convair FW	17-Sep-57	First B-36D conversion. Completed on 5-Oct-50 in Fort Worth. Converted to JB-36D and used for special test programs. To reclamation 7-Nov-57.
44-92055	52	B-36B B-36D B-36D-III	Retired	95th BMW	22-Aug-56	To reclamation 26-Aug-57.
44-92056	53	B-36B B-36D B-36D-II	Retired	92nd BMW	8-Nov-56	To reclamation 28-Mar-57.
44-92057	54	B-36B YB-36D B-36D B-36D-II	Retired	92nd BMW	6-Dec-56	B-36D prototype. First flew 26-Mar-49 with J35 engines. To reclamation 15-Feb-57.
44-92058	55	B-36B B-36D B-36D-II	Retired	92nd BMW	6-Nov-56	To reclamation 28-Mar-57.
44-92059	56	B-36B B-36D B-36D-II	Retired	92nd BMW	14-Dec-56	To reclamation 12-Feb-57.
44-92060	57	B-36B B-36D B-36D-II	Retired	92nd BMW	17-Dec-56	To reclamation 12-Feb-57.
44-92061	58	B-36B B-36D B-36D-III	Retired	42nd BMW	7-Feb-56	To reclamation 9-Jan-57.
44-92062	59	B-36B B-36D B-36D-III	Retired	95th BMW	1-Jun-56	To reclamation 10-Sep-56.
44-92063	60	B-36B B-36D B-36D-III	Retired	95th BMW	27-Apr-56	To reclamation 10-Sep-56.
44-92064	61	B-36B B-36D B-36D-II	Retired	92nd BMW	15-Nov-56	To reclamation 12-Mar-57.
44-92065	62	B-36B B-36D B-36D-II	Retired	92nd BMW	27-Feb-57	To reclamation 27-May-57.

Tail #	CN	Lineage	Status	Last Unit	Date	Comments
44-92066	63	B-36B B-36D B-36D-III	Retired	95th BMW	7-Mar-56	To reclamation 10-Sep-56.
44-92067	64	B-36B B-36D B-36D-II	Retired	92nd BMW	11-Dec-56	To reclamation 12-Mar-57.
44-92068	65	B-36B B-36D B-36D-II	Retired	92nd BMW	20-Dec-56	To reclamation 18-Feb-57.
44-92069	66	B-36B B-36D	Destroyed	92nd BMW	26-Feb-54	Destroyed by fire on the runway at Fairchild AFB, WA. One engine caught fire on run-up and takeoff was aborted. Landing gear then collapsed, and fuel tank ruptured. No fatalities.
44-92070	67	B-36B B-36D B-36D-II	Retired	92nd BMW	5-Mar-57	To reclamation 5-Mar-57.
44-92071	68	B-36B B-36D	Destroyed	7th BMW	11-Dec-53	Damaged at Fairfield-Suisun AFB, CA on 15-Feb-50. Later returned to service. Crashed into mountains near El Paso, TX on 11-Dec-53 during approach due to pilot disorientation. Total 9 fatalities (all crew), 0 survivors.
44-92072	69	B-36B B-36D B-36D-II	Retired	42nd BMW	13-Feb-56	To reclamation 10-Sep-56.
44-92073	70	B-36B B-36D B-36D-III	Retired	92nd BMW	2-Jan-57	To reclamation 18-Feb-57.
44-92074	71	B-36B B-36D B-36D-II	Retired	92nd BMW	6-Dec-56	To reclamation 15-Feb-57.
44-92075	72	B-36B	Destroyed	7th BMW	13-Feb-50	First ever "Broken Arrow" involving a nuclear weapon. Crashed enroute from Alaska to Carswell AFB, TX. Experienced icing conditions and three engines flamed out. Crew abandoned the aircraft, but it continued flying for 200 miles and then crashed into a mountain. Crew jettisoned (unarmed) Mk-4 bomb before bailing out. Total 5 fatalities (all crew), 12 survivors.
44-92076	73	B-36B B-36D B-36D-III	Retired	92nd BMW	26-Nov-56	To reclamation 18-Feb-57.
44-92077	74	B-36B B-36D B-36D-II	Retired	92nd BMW	14-Mar-57	To MASDC via San Antonio - Air Material Area, Kelly AFB, TX. To reclamation 28-Mar-57.
44-92078	75	B-36B B-36D B-36D-II	Retired	92nd BMW	25-Feb-57	To reclamation 27-May-57.

Tail #	CN	Lineage	Status	Last Unit	Date	Comments
44-92079	76	B-36B	Destroyed	7th BMW	15-Sep-49	Crashed after take-off from Carswell AFB, TX. Aircraft crashed into Lake Worth at the end of Carswell runway. Pilot claimed the propellers switched to reverse thrust. Total 5 fatalities (all crew), 8 survivors.
44-92080	77	B-36B B-36D	Destroyed	92nd BMW	29-Jan-52	Landed short, crashed into a snowbank, and burned at Fairchild AFB, WA. No Fatalities.
44-92081	78	B-36B B-36D B-36D-II	Retired	92nd BMW	4-Dec-56	To reclamation 15-Feb-57.
44-92082	79	B-36B B-36D B-36D-III	Retired	92nd BMW	9-Jan-57	To reclamation 18-Feb-57.
44-92083	80	B-36B B-36D B-36D-II	Retired	92nd BMW	13-Dec-56	To reclamation 12-Feb-57.
44-92084	81	B-36B B-36D B-36D-II	Retired	92nd BMW	19-Feb-57	To reclamation 18-Apr-57.
44-92085	82	B-36B B-36D B-36D-III	Retired	92nd BMW	26-Feb-57	To reclamation 17-Jun-57.
44-92086	83	B-36B B-36D B-36D-II	Retired	92nd BMW	20-Dec-56	To reclamation 18-Feb-57.
44-92087	84	B-36B B-36D B-36D-II	Retired	92nd BMW	16-Jan-57	To reclamation 29-Apr-57.
44-92088	85	RB-36D ERB-36D	Retired	SA-ALC		Ordered as B-36B. Converted during production and delivered as RB-36D. Damaged by mechanical failure 3 miles SW of Corning, AR on 12-Aug-50. Later returned to service. Spent its entire service life as a test aircraft. Modified with K-42 Boston Camera and redesignated ERB-36D in 1954. Scrapped at Kelly AFB, TX in late 1955.
44-92089	86	RB-36D RB-36D-III	Retired	72nd BMW	8-Aug-56	Ordered as B-36B. Converted during production and delivered as RB-36D. To reclamation 10-Oct-57.
44-92090	87	RB-36D GRB-36D GRB-36D-III	Retired	AFTC	3-Jul-56	Ordered as B-36B. Converted during production and delivered as RB-36D. Used in accelerated service test to evaluate B-36D operations. Made the longest known B-36 flight lasting 51.5 hours. To MASDC via AFTC Edwards AFB, CA. To reclamation 21-Oct-57.
44-92091	88	RB-36D RB-36D-II	Retired	72nd BMW	26-Aug-56	Ordered as B-36B. Converted during production and delivered as RB-36D. To reclamation 4-Sep-57.

Tail #	CN	Lineage	Status	Last Unit	Date	Comments
44-92092	89	RB-36D GRB-36D GRB-36D-III	Retired	99th BMW	17-Jul-56	Ordered as B-36B. Converted during production and delivered as RB-36D. To reclamation 26-Aug-57.
44-92093	90	RB-36D RB-36D-II	Retired	72nd BMW	11-Sep-56	Ordered as B-36B. Converted during production and delivered as RB-36D. To reclamation 11-Dec-56.
44-92094	91	RB-36D GRB-36D GRB-36D-III	Retired	99th BMW	10-Feb-56	Ordered as B-36B. Converted during production and delivered as RB-36D. To reclamation 30-Sep-57.
44-92095	92	B-36D B-36D-III	Retired	6515th MAG	1-Jun-56	Ordered as B-36B. Converted during production and delivered as B-36D. Scrapped 10-Sep-56.
44-92096	93	B-36D B-36D-III	Retired	95th BMW		Ordered as B-36B. Converted during production and delivered as B-36D. To reclamation 10-Sep-56.
44-92097	94	B-36D	Destroyed	95th BMW	28-Aug-54	Ordered as B-36B. Converted during production and delivered as B-36D. Crashed on approach at Biggs AFB, TX due to loss of engine. Lost altitude and crashed 1,300 feet short of runway and burst into flames. Total 1 fatality (crew).
44-92098	95	B-36D B-36D-II	Retired	92nd BMW	25-Feb-57	Ordered as B-36B. Delivered 18-Aug-50. First delivery of a B-36B converted during production and delivered as B-36D. Scrapped 27-May-57.
49-2647	96	B-36D B-36D-III	Retired	95th BMW	13-Jun-56	First new-build ordered as B-36D. First flew 11-Jul-49. To reclamation 10-Sep-56.
49-2648	97	B-36D B-36D-III	Retired	95th BMW	19-Jun-56	To reclamation 10-Sep-56.
49-2649	98	B-36D B-36D-II	Retired	92nd BMW	27-Mar-56	Reclamation date unknown.
49-2650	99	B-36D B-36D-III	Retired	95th BMW	15-Jun-56	Reclamation date unknown.
49-2651	101	B-36D B-36D-III	Retired	95th BMW	6-Jul-56	To reclamation 12-Mar-57.
49-2652	102	B-36D B-36D-III	Retired	92nd BMW	23-Jan-57	To reclamation 17-Jun-57.
49-2653	103	B-36D B-36D-III	Destroyed	SWC KAFB	27-Jun-55	Delivered 22-Aug-50. First delivery of a new-build ordered as B-36D. Damaged in tornado at Carswell AFB, TX on 1-Sep-52. Later returned to service. Participated in the first B-36D gunnery mission on 12-Sep-50. Participated in atomic test Operations IVY, UPSHOT-KNOTHOLE, and CASTLE. Received structural damage during CASTLE in 1954. Scrapped 27-Jun-55.
49-2654	104	B-36D B-36D-III	Retired	95th BMW	4-Sep-56	To reclamation 23 Sep-56.
49-2655	135	B-36D B-36D-III	Retired	95th BMW	27-Aug-56	To reclamation 13-Sep-57.

B-36 Peacemaker: The Big Stick of Strategic Air Command

Tail #	CN	Lineage	Status	Last Unit	Date	Comments
49-2656	110	B-36D B-36D-III	Retired	95th BMW	11-Jul-56	To reclamation 12-Mar-57.
49-2657	111	B-36D B-36D-II	Retired	42nd BMW	15-Feb-56	Reclamation date unknown.
49-2658	115	B-36D	Destroyed	7th BMW	27-Apr-51	Took off from Carswell AFB, TX and crashed near Perkins, OK after mid-air collision with P-51D Mustang during gunnery training. Total 13 fatalities (12 crew/1 P-51 pilot), 4 survivors.
49-2659	116	B-36D B-36D-II	Retired	42nd BMW	17-Feb-56	To reclamation 01-Jun-56.
49-2660	117	B-36D	Destroyed	7th BMW	6-May-51	Crashed during landing in high winds at Kirtland AFB, NM. Pilot unable to keep wings level and No. 6 propeller hit the ground and caused engine fire. Pilot attempted go-around before the aircraft crashed. Total 23 fatalities, 2 survivors.
49-2661	121	B-36D	Destroyed	7th BMW	5-Aug-52	Crashed in the ocean off the coast near San Diego. Fire on No. 4 engine spread to the wing. Crew bailed out. Total 2 fatalities (all crew), 6 survivors.
49-2662	122	B-36D B-36D-II	Retired	92nd BMW	15-Mar-57	To reclamation 17-Jun-57.
49-2663	123	B-36D B-36D-III	Retired	95th BMW	10-Sep-56	Reclamation date unknown.
49-2664	127	B-36D B-36D-III	Retired	95th BMW	1-Jun-56	To reclamation 10-Sep-56.
49-2665	128	B-36D B-36D-III	Retired	95th BMW	03-Aug-56	To reclamation 10-Oct-57.
49-2666	129	B-36D B-36D-II	Retired	92nd BMW	26-Nov-56	To reclamation 15-Feb-57.
49-2667	133	B-36D B-36D-III	Retired	95th BMW	31-Aug-56	To reclamation 13-Sep-57.
49-2668	134	B-36D B-36D-II	Retired	42nd BMW	20-Mar-56	To reclamation 10-Sep-56.
49-2669	109	YB-36F B-36F B-36F-III	Retired	6th BMW	19-Aug-57	B-36F prototype. First flew 18-Nov-50. To reclamation 7-Nov-57.
49-2670	139	B-36F B-36F-II	Retired	6th BMW	26-Jun-57	Part of accelerated test program in Fort Worth before delivery. To reclamation 5-Aug-57.
49-2671	140	B-36F B-36F-II	Retired	6th BMW	19-Jun-57	First B-36F delivered to SAC on 18-Aug-51. Part of accelerated test program in Fort Worth before delivery. To reclamation 28-Jun-57.
49-2672	141	B-36F B-36F-II	Retired	6th BMW	3-Jun-57	Part of accelerated test program in Fort Worth before delivery. To reclamation 4-Jun-57.
49-2673	145	B-36F B-36F-II	Retired	6th BMW	26-Jun-57	To reclamation 8-Jul-57.

Tail #	CN	Lineage	Status	Last Unit	Date	Comments
49-2674	146	B-36F B-36F-II	Retired	6th BMW	10-Jun-57	To reclamation 10-Jun-57.
49-2675	147	B-36F B-36F-II	Retired	6th BMW	17-Jun-57	To reclamation 17-Jun-57.
49-2676	151	B-36G YB-60	Cancelled	Convair FW	Jul-54	Originally ordered as B-36F. Delivered as B-36G and then redesignated as YB-60. First prototype for all jet, swept wing variant designed to compete with the B-52. Good challenger, but approximately 100 mph slower. Program cancelled 20-Jan-53. Scrapped Jul-54.
49-2677	152	B-36F JB-36F JB-36F-II	Retired	Convair FW	19-Aug-57	Converted to JB-36F and used for special test programs. It was used to fly a B-58 static test aircraft from Fort Worth to Wright-Patterson AFB, OH. To reclamation 7-Nov-57.
49-2678	153	B-36F B-36F-II	Retired	6th BMW	20-Jun-57	To reclamation 29-Jul-57.
49-2679	157	B-36F B-36F-II	Destroyed	7th BMW	4-Aug-52	Destroyed by fire on flightline at Carswell AFB, TX due overflow from No. 3 fuel tank vent that was ignited by nearby ground power unit exhaust. No Fatalities.
49-2680	158	B-36F B-36F-II	Retired	6th BMW	24-Jun-57	To reclamation 5-Aug-57.
49-2681	159	B-36F B-36F-II	Retired	6th BMW	3-Jun-57	To reclamation 4-Jun-57.
49-2682	163	B-36F B-36F-II	Retired	6th BMW	18-Jun-57	To reclamation 28-Jun-57.
49-2683	164	B-36F B-36F-II	Retired	6th BMW	4-Jun-57	To reclamation 4-Jun-57.
49-2684	165	B-36G YB-60	Cancelled	Convair FW	Jul-54	Originally ordered as B-36F. Delivered as B-36G and then redesignated as YB-60. Second prototype for all jet, swept wing variant designed to compete with the B-52. Good challenger, but approximately 100 mph slower. Program cancelled 20-Jan-53. Never flew, scrapped Jul-54.
49-2685	170	B-36F B-36F-II	Retired	6th BMW	1-Jul-57	To reclamation 16-Aug-57.
49-2686	100	RB-36D RB-36D-II	Retired	72nd BMW	12-Aug-56	To reclamation 30-Sep-57.
49-2687	105	RB-36D GRB-36D GRB-36D-III	Retired	99th BMW	10-Jul-56	To reclamation 13-Sep-57.
49-2688	106	RB-36D RB-36D-II	Retired	72nd BMW	14-Sep-56	To reclamation 11-Dec-56.
49-2689	107	RB-36D RB-36D-II	Retired	72nd BMW	7-Aug-56	To reclamation 23-Sep-57.
49-2690	108	RB-36D RB-36D-II	Retired	72nd BMW	25-Jun-56	To reclamation 12-Mar-57.

Tail #	CN	Lineage	Status	Last Unit	Date	Comments
49-2691	112	RB-36D RB-36D-II	Retired	72nd BMW	11-Sep-56	To reclamation 15-Feb-57.
49-2692	113	RB-36D GRB-36D GRB-36D-III	Retired	99th BMW	24-Jul-56	To reclamation 23-Sep-57.
49-2693	114	RB-36D RB-36D-II	Retired	72nd BMW	25-Sep-56	To reclamation 12-Mar-57.
49-2694	118	RB-36D GRB-36D GRB-36D-III	Retired	6515th MG	1-Jun-56	To reclamation 21-Oct-57.
49-2695	119	RB-36D GRB-36D GRB-36D-III	Retired	99th BMW	24-Jul-56	To reclamation 4-Sep-57.
49-2696	120	RB-36D GRB-36D GRB-36D-III	Retired	Convair FW	24-May-56	To reclamation 21-Oct-57.
49-2697	124	RB-36D RB-36D-II	Retired	72nd BMW	21-Sep-56	To reclamation 18-Feb-57.
49-2698	125	RB-36D RB-36D-II	Retired	72nd BMW	4-Sep-56	To reclamation 11-Dec-56.
49-2699	126	RB-36D RB-36D-II	Retired	72nd BMW	12-Aug-56	To reclamation 4-Sep-57.
49-2700	130	RB-36D RB-36D-II	Retired	72nd BMW	5-Aug-56	To reclamation 30-Sep-57.
49-2701	131	RB-36D GRB-36D GRB-36D-III	Retired	99th BMW	10-Jul-56	To reclamation 23-Sep-57.
49-2702	132	RB-36D GRB-36D GRB-36D-III	Retired	99th BMW	17-Jul-56	To reclamation 26-Aug-57.
49-2703	136	RB-36F RB-36F-II	Retired	72nd BMW	1-Oct-58	First RB-36F assigned to SAC (5th SRW) on 28-May-51. Part of accelerated test program in Fort Worth before delivery. To reclamation 2-Oct-58.
49-2704	137	RB-36F RB-36F-II	Retired	72nd BMW	22-Oct-58	Part of accelerated test program in Fort Worth before delivery. To reclamation 24-Oct-58.
49-2705	138	RB-36F RB-36F-II	Retired	72nd BMW	8-Oct-58	Part of accelerated test program in Fort Worth. To reclamation 9-Oct-58.
49-2706	142	RB-36F RB-36F-II	Retired	72nd BMW	3-Jul-58	To reclamation 7-Jul-58.
49-2707	143	RB-36F JRB-36F	Retired	Convair FW	9-May-57	FICON prototype. First flew Jan-52. Re-designated JRB-36F on 30-Oct-56. To reclamation 16-Aug-57.
49-2708	144	RB-36F RB-36F-II	Retired	72nd BMW	17-Jun-58	To reclamation 18-Jun-58.
49-2709	148	RB-36F RB-36F-II	Retired	72nd BMW	11-Dec-57	To reclamation 16-Dec-57.
49-2710	149	RB-36F RB-36F-II	Retired	72nd BMW	1-Jul-58	To reclamation 2-Jul-58.

Tail #	CN	Lineage	Status	Last Unit	Date	Comments
49-2711	150	RB-36F RB-36F-II	Retired	72nd BMW	22-Jul-58	To reclamation 23-Jul-58.
49-2712	154	RB-36F RB-36F-II	Retired	72nd BMW	28-Oct-58	To reclamation 29-Oct-58.
49-2713	155	RB-36F RB-36F-II	Retired	72nd BMW	24-Jun-58	To reclamation 25-Jun-58.
49-2714	156	RB-36F RB-36F-II	Retired	72nd BMW	10-Jul-58	To reclamation 11-Jul-58.
49-2715	160	RB-36F RB-36F-II	Retired	72nd BMW	7-Oct-58	To reclamation 8-Oct-58.
49-2716	161	RB-36F RB-36F-II	Retired	72nd BMW	26-Feb-58	To reclamation 27-Feb-58.
49-2717	162	RB-36F RB-36F-II	Retired	72nd BMW	23-Jul-58	To reclamation 25-Jul-58.
49-2718	166	RB-36F RB-36F-II	Retired	72nd BMW	2-Jul-58	To reclamation 7-Jul-58.
49-2719	167	RB-36F RB-36F-II	Retired	72nd BMW	10-Jul-58	To reclamation 11-Jul-58.
49-2720	168	RB-36F RB-36F-II	Retired	72nd BMW	31-Jul-58	To reclamation 4-Aug-58.
49-2721	169	RB-36F RB-36F-II	Retired	72nd BMW	3-Jul-58	To reclamation 7-Jul-58.
50-1064	171	B-36F B-36F-III	Retired	6th BMW	6-Aug-57	To reclamation 7-Nov-57.
50-1065	172	B-36F B-36F-III	Retired	6th BMW	27-Aug-57	To reclamation 7-Nov-57.
50-1066	173	B-36F	Destroyed	11th BMW	28-May-52	Crashed on landing at Carswell AFB, TX. Engine cowling came off during take-off. Pilot went around and attempted overweight landing. Gear collapsed and aircraft skidded down runway and caught fire. Total 7 fatalities (all crew).
50-1067	175	B-36F	Destroyed	7th BMW	6-Mar-52	Crashed on landing at Carswell AFB, TX. Landing gear collapsed resulting in fuel leak and fire. All crew escaped. No fatalities.
50-1068	176	B-36F B-36F-II	Retired	6th BMW	3-Jul-57	To reclamation 7-Nov-57.
50-1069	177	B-36F B-36F-II	Retired	6th BMW	12-Jun-57	To reclamation 12-Jun-57.
50-1070	178	B-36F B-36F-II	Retired	6th BMW	24-Jun-57	To reclamation 5-Aug-57.
50-1071	180	B-36F B-36F-II	Retired	6th BMW	21-Aug-57	To reclamation 7-Nov-57.
50-1072	181	B-36F B-36F-II	Retired	6th BMW	16-Aug-57	To reclamation 7-Nov-57.
50-1073	182	B-36F B-36F-II	Retired	6th BMW	9-Jul-57	To reclamation 7-Nov-57.
50-1074	183	B-36F B-36F-II	Retired	6th BMW	23-Aug-57	To reclamation 7-Nov-57.

Tail #	CN	Lineage	Status	Last Unit	Date	Comments
50-1075	185	B-36F B-36F-II	Retired	6th BMW	26-Aug-57	To reclamation 7-Nov-57.
50-1076	186	B-36F B-36F-II	Retired	6th BMW	13-Jun-57	To reclamation 13-Jun-57.
50-1077	187	B-36F B-36F-II	Retired	6th BMW	7-Jul-57	To reclamation 7-Nov-57.
50-1078	188	B-36F B-36F-II	Retired	6th BMW	1-Jul-57	To reclamation 7-Nov-57.
50-1079	190	B-36F B-36F-II	Retired	6th BMW	2-Jul-57	To reclamation 7-Nov-57.
50-1080	191	B-36F B-36F-II	Retired	6th BMW	8-Aug-57	To reclamation 7-Nov-57.
50-1081	192	B-36F B-36F-III	Retired	6th BMW	12-Aug-57	To reclamation 7-Nov-57.
50-1082	193	B-36F B-36F-II	Retired	6th BMW	14-Aug-57	To reclamation 7-Nov-57.
50-1083	195	YB-36H B-36H B-36H-III	Retired	7th BMW	23-Jan-58	B-36H prototype. First flew 5-Apr-52. Participated in Operation CASTLE in 1954. To reclamation 26-Mar-58.
50-1084	196	B-36H B-36H-III	Retired	95th BMW	19-Feb-58	To reclamation 21-Feb-58.
50-1085	197	B-36H DB-36H EDB-36H EDB-36H-II	Retired	APG	3-Jun-56	Only production DB-36H converted before the program cancelled. To MASDC via APG Eglin AFB, FL. To reclamation 27-Dec-56.
50-1086	198	B-36H B-36H-III	Retired	7th BMW	10-Apr-58	Participated in CASTLE nuclear tests at Eniwetok and Bikini Atolls in 1954. To reclamation 15-Apr-58.
50-1087	200	B-36H B-36H-II	Retired	95th BMW	11-Feb-58	To reclamation 21-Feb-58.
50-1088	201	B-36H B-36H-II	Retired	95th BMW	10-Jul-57	To reclamation 7-Nov-57.
50-1089	202	B-36H B-36H-II	Retired	95th BMW	27-Feb-58	To reclamation 3-Mar-58.
50-1090	204	B-36H B-36H-II	Retired	95th BMW	13-Feb-58	To reclamation 15-Apr-58.
50-1091	205	B-36H B-36H-III	Retired	95th BMW	12-Jul-57	To reclamation 7-Nov-57.
50-1092	207	B-36H B-36H-II	Retired	95th BMW	2-Aug-57	To reclamation 7-Nov-57.
50-1093	208	B-36H B-36H-III	Retired	7th BMW	16-Jan-58	To reclamation 17-Jan-58.
50-1094	210	B-36H B-36H-III	Retired	95th BMW	11-Jul-57	To reclamation 7-Nov-57.
50-1095	211	B-36H B-36H-II	Retired	95th BMW	5-Aug-57	To reclamation 7-Nov-57.
50-1096	213	B-36H B-36H-II	Retired	95th BMW	18-Mar-57	To reclamation 5-Apr-57.
50-1097	214	B-36H B-36H-II	Retired	95th BMW	21-Aug-57	To reclamation 7-Nov-57.

Tail #	CN	Lineage	Status	Last Unit	Date	Comments
50-1098	174	RB-36F RB-36F-II	Retired	72nd BMW	8-Jul-58	To reclamation 9-Jul-58.
50-1099	179	RB-36F RB-36F-II	Retired	72nd BMW	9-Jul-58	To reclamation 10-Jul-58.
50-1100	184	RB-36F RB-36F-II	Retired	72nd BMW	17-Jul-58	To reclamation 21-Jul-58.
50-1101	189	RB-36F RB-36F-III	Retired	72nd BMW	21-Oct-58	To reclamation 22-Oct-58.
50-1102	194	RB-36F RB-36F-II	Retired	72nd BMW	30-Jul-58	To reclamation 31-Jul-58.
50-1103	199	RB-36H RB-36H-III	Retired	5th BMW	25-Sep-58	To reclamation 29-Sep-58.
50-1104	203	RB-36H RB-36H-III	Retired	5th BMW	10-Jun-58	First RB-36H delivered to SAC (5th SRW) on 7-Feb-52. To reclamation 11-Jun-58.
50-1105	206	RB-36H RB-36H-III	Retired	5th BMW	24-Jun-58	To reclamation 25-Jun-58.
50-1106	209	RB-36H RB-36H-III	Retired	5th BMW	4-Jun-58	To reclamation 5-Jun-58.
50-1107	212	RB-36H RB-36H-III	Retired	5th BMW	5-Jun-58	To reclamation 9-Jun-58.
50-1108	215	RB-36H RB-36H-III	Retired	5th BMW	24-Sep-58	To reclamation 25-Sep-58.
50-1109	216	RB-36H RB-36H-III	Retired	5th BMW	24-Sep-58	To reclamation 25-Sep-58.
50-1110	217	RB-36H RB-36H-III	Retired	5th BMW	30-Sep-58	To reclamation 2-Oct-58.
51-13717	249	RB-36H RB-36H-II	Retired	72nd BMW	28-Sep-58	To reclamation 2-Oct-58.
51-13718	251	RB-36H RB-36H-II	Retired	28th BMW	21-Mar-57	To reclamation 5-Apr-57.
51-13719	253	RB-36H	Destroyed	28th BMW	18-Feb-53	Crashed on landing at Walker AFB, NM. Landing gear collapsed resulting in a ruptured fuel tank and fire on the runway. All crew escaped. No fatalities, 22 survivors.
51-13720	255	RB-36H RB-36H-II	Destroyed	28th BMW	15-Nov-56	Crashed after take-off from Lowry AFB, CO. Loss of power due to fuel starvation. Pilot attempted landing at Denver Stapleton Airport but crashed 1 mile short of the runway and was destroyed by fire. No fatalities, 21 survivors.
51-13721	257	RB-36H	Destroyed	28th SRW	18-Mar-53	Crashed near Burgoynes Cove, Newfoundland, Canada after going off course in bad weather. Aircraft was enroute from Lajes Airdrome, Azores to home station Rapid City AFB, SD when it crashed into an 896-foot ridge, 30 miles inland. Total 23 fatalities (all crew), 0 survivors (fatalities included BG Richard Ellsworth). Also, an SB-29 search plane crashed with 10 fatalities, 0 survivors.

Tail #	CN	Lineage	Status	Last Unit	Date	Comments
51-13722	259	RB-36H	Destroyed	28th SRW	27-Aug-54	Crashed at the end of the runway after performing Planned Position Indicator Ground Control Approaches (GCA). The aircraft hit an inoperative obstruction light and crashed. The GCA radar was miscalibrated by 1/2 mile and caused a 150 feet low glideslope indication. Total 26 fatalities (all crew), 1 survivor (3 crew survived the crash but 2 died within 1 week of crash).
51-13723	261	RB-36H RB-36H-III	Retired	72nd BMW		To reclamation Nov-58.
51-13724	263	RB-36H RB-36H-II	Retired	28th BMW	21-May-57	To reclamation 4-Jun-57.
51-13725	265	RB-36H RB-36H-II	Retired	28th BMW	26-Mar-57	To reclamation 13-May-57.
51-13726	267	RB-36H RB-36H-II	Retired	28th BMW	1-May-57	To reclamation 13-May-57.
51-13727	269	RB-36H RB-36H-II	Retired	72nd BMW	12-Nov-58	To reclamation 18-Nov-58.
51-13728	271	RB-36H RB-36H-II	Retired	28th BMW	15-May-57	To reclamation 17-Jul-57.
51-13729	273	RB-36H RB-36H-II	Retired	28th BMW	6-May-57	To reclamation 17-Jul-57.
51-13730	275	RB-36H RB-36H-II	Display	28th BMW	1-Mar-57	Transferred to Chanute AFB, IL as ground trainer in Mar-57. Put on static display at Chanute under Air Force Museum loan program in 1970. Moved to Castle AFB, CA Museum and put on display in 1994.
51-13731	277	RB-36H RB-36H-II	Retired	28th BMW	23-May-57	To reclamation 6-Jun-57.
51-13732	279	RB-36H RB-36H-II	Retired	28th BMW	22-May-57	To reclamation 4-Jun-57.
51-13733	281	RB-36H RB-36H-II	Retired	28th BMW	16-May-57	To reclamation 6-Jun-57.
51-13734	283	RB-36H RB-36H-III	Retired	28th BMW	13-May-57	To reclamation 17-Jul-57.
51-13735	285	RB-36H RB-36H-III	Retired	5th BMW	18-Jun-58	To reclamation 19-Jun-58.
51-13736	287	RB-36H RB-36H-II	Retired	5th BMW	12-Jun-58	To reclamation 13-Jun-58.
51-13737	289	RB-36H RB-36H-II	Retired	28th BMW	6-May-57	To reclamation 17-Jul-57.
51-13738	291	RB-36H RB-36H-III	Retired	5th BMW	25-Jun-58	To reclamation 26-Jun-58.
51-13739	293	RB-36H RB-36H-III	Retired	5th BMW	25-Jun-58	To reclamation 26-Jun-58.
51-13740	295	RB-36H RB-36H-II	Retired	5th BMW	11-Jun-58	To reclamation 13-Jun-58.
51-13741	297	RB-36H RB-36H-II	Retired	5th BMW	3-Jun-58	To reclamation 4-Jun-58.

Tail #	CN	Lineage	Status	Last Unit	Date	Comments
51-5699	218	B-36H B-36H-III	Retired	95th BMW	10-Jun-57	To reclamation 10-Jun-57.
51-5700	219	B-36H B-36H-III	Retired	7th BMW	14-May-58	To reclamation 19-May-58.
51-5701	220	B-36H B-36H-III	Retired	11th BMW	19-Nov-57	To reclamation 21-Nov-57.
51-5702	222	B-36H B-36H-III	Retired	11th BMW	9-Oct-57	To reclamation 7-Nov-57.
51-5703	224	B-36H B-36H-III	Retired	11th BMW	29-Aug-57	To reclamation 7-Nov-57.
51-5704	226	B-36H B-36H-III	Retired	7th BMW	2-Jun-58	To reclamation 3-Jun-58.
51-5705	228	B-36H B-36H-II	Retired	95th BMW	27-Aug-57	To reclamation 7-Nov-57.
51-5706	230	B-36H DB-36H EDB-36H JDB-36H JDB-36H-II	Retired	Bell HAFB	20-Sep-57	Tanbo XIV prototype. Transferred to DB-36H program on 6-Jul-54. To reclamation 7-Nov-57.
51-5707	232	B-36H B-36H-II	Retired	95th BMW	6-Jun-57	To reclamation 6-Jun-57.
51-5708	234	B-36H B-36H-III	Retired	11th BMW	12-Sep-57	To reclamation 7-Nov-57.
51-5709	236	B-36H B-36H-II	Retired	95th BMW	7-Jun-57	To reclamation 7-Jun-57.
51-5710	238	B-36H YDB-36H DB-36H EDB-36H JDB-36H JDB-36H-II	Retired	Bell HAFB	16-Jun-57	DB-36H prototype. First flew 3-Jul-53. To reclamation 5-Sep-57.
51-5711	240	B-36H B-36H-III	Retired	11th BMW	12-Nov-57	To reclamation 12-Nov-57.
51-5712	242	B-36H XB-36H NB-36H	Retired	Convair FW	1-Sep-58	Damaged in tornado at Carswell AFB, TX on 1-Sep-52. Assigned to Project MX-1589 and modified as Nuclear Test Aircraft in 1953. Redesignated XB-36H on 11-Mar-55. Redesignated NB-36H on 6-Jun-56. Salvaged at Convair-FW in Sep-58.
51-5713	244	B-36H B-36H-III	Retired	7th BMW	7-Feb-58	To reclamation 10-Feb-58.
51-5714	246	B-36H B-36H-III	Retired	7th BMW	28-Jan-58	To reclamation 29-Jan-58.
51-5715	248	B-36H B-36H-III	Retired	11th BMW	18-Sep-57	To reclamation 7-Nov-57.
51-5716	250	B-36H B-36H-III	Retired	7th BMW	21-May-58	To reclamation 23-May-58.
51-5717	252	B-36H B-36H-III	Retired	7th BMW	2-May-58	To reclamation 6-May-58.
51-5718	254	B-36H	Retired	11th BMW	14-Nov-57	To reclamation 21-Nov-57.

Tail #	CN	Lineage	Status	Last Unit	Date	Comments
		B-36H-III				
51-5719	256	B-36H	Destroyed	7th BMW	7-Feb-53	Crashed in Nethermore Woods after flying two missed approaches at RAF Fairford, UK. Ran short of fuel and crew abandoned the aircraft. It flew an additional 30 miles before crashing. No fatalities, 14 survivors.
51-5720	258	B-36H B-36H-III	Retired	7th BMW	14-Nov-57	Reclamation date unknown.
51-5721	260	B-36H B-36H-III	Retired	7th BMW	18-Feb-58	To reclamation 19-Feb-58.
51-5722	262	B-36H B-36H-III	Retired	11th BMW	20-Nov-57	To reclamation 22-Nov-57.
51-5723	264	B-36H B-36H-III	Retired	11th BMW	13-Aug-57	To reclamation 7-Nov-57.
51-5724	266	B-36H B-36H-III	Retired	11th BMW	4-Sep-57	To reclamation 7-Nov-57.
51-5725	268	B-36H B-36H-III	Retired	7th BMW	30-Jan-58	To reclamation 31-Jan-58.
51-5726	270	B-36H EB-36H JB-36H JB-36H-III	Retired	SWC KAFB	19-Jun-57	Participated in multiple atomic test operations. Also, used to measure cosmic ray neutrons in the upper atmosphere under the COSMIC RAYS investigations. To reclamation 16-Aug-57.
51-5727	272	B-36H B-36H-III	Retired	11th BMW	2-Oct-57	To reclamation 7-Nov-57.
51-5728	274	B-36H B-36H-III	Retired	11th BMW	10-Sep-57	To reclamation 7-Nov-57.
51-5729	276	B-36H	Destroyed	7th BMW	12-Feb-53	Crashed on a hill at Goose Bay, Labrador on return to Carswell AFB, TX from the UK. Aircraft ran out of fuel while in the pattern waiting to land. Total 2 fatalities (all crew), 15 survivors.
51-5730	278	B-36H B-36H-III	Retired	7th BMW	24-Feb-58	To reclamation 25-Feb-58.
51-5731	280	B-36H EB-36H JB-36H JB-36H-III	Retired	SWC KAFB	18-Mar-57	Participated in multiple atomic test operations. To reclamation 5-Apr-57.
51-5732	282	B-36H B-36H-III	Retired	11th BMW	16-Oct-57	To reclamation 7-Nov-57.
51-5733	284	B-36H B-36H-III	Retired	7th BMW	7-Jan-58	To reclamation 8-Jan-58.
51-5734	286	B-36H B-36H-III	Retired	11th BMW	27-Aug-57	Used for startup and take-off sequences in the movie "Strategic Air Command" starring Jimmy Stewart. To reclamation 7-Nov-57.
51-5735	288	B-36H B-36H-III	Retired	11th BMW	25-Sep-57	To reclamation 7-Nov-57.
51-5736	290	B-36H B-36H-III	Retired	11th BMW	21-Aug-57	To reclamation 7-Nov-57.

Tail #	CN	Lineage	Status	Last Unit	Date	Comments
51-5737	292	B-36H B-36H-III	Retired	7th BMW	9-Jan-58	To reclamation 9-Jan-58.
51-5738	294	B-36H B-36H-III	Retired	11th BMW	14-Aug-57	To reclamation 7-Nov-57.
51-5739	296	B-36H B-36H-III	Retired	11th BMW	28-Aug-57	To reclamation 7-Nov-57.
51-5740	298	B-36H B-36H-III	Retired	11th BMW	15-Aug-57	To reclamation 7-Nov-57.
51-5741	299	B-36H B-36H-III	Destroyed	7th BMW	7-Jun-57	Damaged in thunderstorm near Carswell AFB, TX on 25-Apr-57. Written off and used for fire fighter training. Scrapped 7-Jun-57.
51-5742	300	B-36H B-36H-III	Retired	7th BMW	17-Apr-58	To reclamation 21-Apr-58.
51-5743	221	RB-36H RB-36H-III	Retired	72nd BMW	5-Nov-58	To reclamation 6-Nov-58.
51-5744	223	RB-36H RB-36H-III	Retired	72nd BMW	9-Oct-58	To reclamation 9-Oct-58.
51-5745	225	RB-36H RB-36H-III	Destroyed	72nd BMW	9-Nov-57	Forced to land with engine failure at Andersen AFB, Guam on 6-May-55. Destroyed by fire at Ramey AFB, Puerto Rico on 9-Nov-57. No fatalities.
51-5746	227	RB-36H RB-36H-III	Retired	72nd BMW	13-Nov-58	To reclamation 14-Nov-58.
51-5747	229	RB-36H RB-36H-III	Retired	72nd BMW	6-Nov-58	To reclamation 10-Nov-58.
51-5748	231	RB-36H JRB-36H JRB-36H-III	Retired	SWC KAFB	20-Aug-58	Converted to JRB-36H and modified with upward looking camera to photograph mushroom clouds during Operation HARDTRACK I atomic tests. To reclamation 22-Aug-58.
51-5749	233	RB-36H RB-36H-II	Retired	28th BMW	28-Mar-57	To reclamation 13-May-57.
51-5750	235	RB-36H JRB-36H JRB-36H-II	Retired	SWC KAFB	20-Aug-58	Converted to JRB-36H and modified with upward looking camera to photograph mushroom clouds during Operation HARDTRACK I atomic tests. To reclamation 22-Aug-58.
51-5751	237	RB-36H RB-36H-II	Retired	28th BMW	26-Mar-57	To reclamation 13-May-57.
51-5752	239	RB-36H RB-36H-II	Retired	28th BMW	21-Mar-57	To reclamation 6-Apr-57.
51-5753	241	RB-36H RB-36H-II	Retired	72nd BMW	18-Nov-58	To reclamation 19-Nov-58.
51-5754	243	RB-36H RB-36H-II	Retired	72nd BMW	4-Nov-58	To reclamation 6-Nov-58.
51-5755	245	RB-36H RB-36H-II	Retired	72nd BMW	12-Nov-58	To reclamation 13-Nov-58.
51-5756	247	RB-36H RB-36H-II	Retired	72nd BMW	12-Nov-58	To reclamation 13-Nov-58.

Tail #	CN	Lineage	Status	Last Unit	Date	Comments
52-1343	304	B-36H B-36H-III	Retired	7th BMW	8-Apr-58	To reclamation 15-Apr-58.
52-1344	306	B-36H B-36H-III	Retired	7th BMW	24-Apr-58	To reclamation 25-Apr-58.
52-1345	308	B-36H B-36H-III	Retired	11th BMW	7-Aug-57	To reclamation 7-Nov-57.
52-1346	310	B-36H B-36H-III	Retired	7th BMW	5-Feb-58	To reclamation 15-Apr-58.
52-1347	312	B-36H B-36H-III	Retired	7th BMW	6-May-58	To reclamation 6-May-58.
52-1348	314	B-36H B-36H-III	Retired	7th BMW	8-May-58	To reclamation 9-May-58.
52-1349	316	B-36H B-36H-III	Retired	7th BMW	21-Jan-58	To reclamation 22-Jan-58.
52-1350	318	B-36H B-36H-III	Retired	7th BMW	15-Apr-58	To reclamation 17-Apr-58.
52-1351	320	B-36H B-36H-III	Retired	11th BMW	9-Aug-57	To reclamation 7-Nov-57.
52-1352	322	B-36H B-36H-III	Retired	11th BMW	19-Aug-57	To reclamation 7-Nov-57.
52-1353	324	B-36H B-36H-III	Retired	11th BMW	6-Nov-57	To reclamation 7-Nov-57.
52-1354	326	B-36H B-36H-III	Retired	11th BMW	5-Aug-57	To reclamation 7-Nov-57.
52-1355	328	B-36H B-36H-III	Retired	7th BMW	25-Feb-58	To reclamation 29-Feb-58.
52-1356	330	B-36H B-36H-III	Retired	11th BMW	23-Oct-57	To reclamation 7-Nov-57.
52-1357	332	B-36H EB-36H JB-36H	Retired	SWC KAFB	3-Jan-58	Participated in multiple atomic test operations. To reclamation 6-Jan-58.
52-1358	334	B-36H EB-36H JB-36H	Retired	SWC KAFB	29-Apr-57	Participated in multiple atomic test operations. To reclamation 16-Aug-57.
52-1359	336	B-36H B-36H-III	Retired	7th BMW	14-Jan-58	To reclamation 15-Jan-58.
52-1360	338	B-36H B-36H-III	Retired	7th BMW	22-Apr-58	To reclamation 22-Apr-58.
52-1361	340	B-36H B-36H-III	Retired	11th BMW	20-Aug-57	To reclamation 7-Nov-57.
52-1362	342	B-36H B-36H-III	Retired	7th BMW	2-Apr-58	To reclamation 3-Apr-58.
52-1363	344	B-36H B-36H-III	Retired	11th BMW	22-Aug-57	To reclamation 7-Nov-57.
52-1364	346	B-36H B-36H-III	Retired	11th BMW	26-Aug-57	To reclamation 7-Nov-57.
52-1365	348	B-36H B-36H-III	Retired	11th BMW	1-Jul-57	To reclamation 7-Nov-57.
52-1366	350	B-36H B-36H-III	Retired	7th BMW	27-Feb-58	To reclamation 3-Mar-58.

B-36 PEACEMAKER: THE BIG STICK OF STRATEGIC AIR COMMAND

Tail #	CN	Lineage	Status	Last Unit	Date	Comments
52-1367	301	RB-36H RB-36H-II	Retired	5th BMW	19-Jun-58	To reclamation 20-Jun-58.
52-1368	302	RB-36H RB-36H-II	Retired	5th BMW	17-Jun-58	To reclamation 19-Jun-58.
52-1369	303	RB-36H RB-36H-II	Destroyed	5th BMW	5-Aug-53	Crashed west of Scotland due to an engine fire during flight from Travis AFB, CA to RAF Lakenheath, UK. Total 19 fatalities (all crew), 4 survivors.
52-1370	305	RB-36H RB-36H-II	Retired	5th BMW	28-Aug-58	To reclamation 29-Aug-58.
52-1371	307	RB-36H RB-36H-II	Retired	5th BMW	17-Jul-58	To reclamation 21-Jul-58.
52-1372	309	RB-36H RB-36H-II	Retired	5th BMW	21-Aug-58	To reclamation 22-Aug-58.
52-1373	311	RB-36H RB-36H-II	Retired	5th BMW	20-May-58	To reclamation 6-Jun-58.
52-1374	313	RB-36H RB-36H-II	Retired	5th BMW	15-Jul-58	To reclamation 17-Jul-58.
52-1375	315	RB-36H RB-36H-II	Retired	5th BMW	20-May-57	To reclamation 29-Jul-57.
52-1376	317	RB-36H RB-36H-II	Retired	5th BMW	27-Aug-58	To reclamation 28-Aug-58.
52-1377	319	RB-36H RB-36H-II	Retired	5th BMW	20-May-57	To reclamation 29-Jul-57.
52-1378	321	RB-36H RB-36H-II	Retired	5th BMW	19-Aug-58	To reclamation 20-Aug-58.
52-1379	323	RB-36H RB-36H-II	Retired	5th BMW	27-May-57	To reclamation 6-Jun-57.
52-1380	325	RB-36H RB-36H-II	Retired	5th BMW	20-Aug-58	To reclamation 22-Aug-58.
52-1381	327	RB-36H RB-36H-II	Retired	5th BMW	26-Aug-58	To reclamation 27-Aug-58.
52-1382	329	RB-36H RB-36H-II	Retired	5th BMW	11-Jun-58	To reclamation 13-Jun-58.
52-1383	331	RB-36H RB-36H-II	Retired	5th BMW	18-Jun-58	To reclamation 19-Jun-58.
52-1384	333	RB-36H RB-36H-II	Retired	5th BMW	26-Jun-58	To reclamation 27-Jun-58.
52-1385	335	RB-36H RB-36H-II	Retired	5th BMW	25-Sep-58	To reclamation 29-Sep-58.
52-1386	337	RB-36H RB-36H-III	Retired	28th BMW	29-May-57	Participated in Operation CASTLE nuclear tests at Eniwetok and Bikini Atolls in 1954. To reclamation 25-Jun-57.
52-1387	339	RB-36H RB-36H-III	Destroyed	28th SRW	4-Jan-56	Crashed during landing at Ellsworth AFB, SD due to uneven power reduction during landing. No fatalities.
52-1388	341	RB-36H RB-36H-II	Retired	5th BMW	19-Jun-58	To reclamation 25-Jun-58.
52-1389	343	RB-36H RB-36H-III	Retired	28th BMW	28-May-57	To reclamation 24-Jun-57.

Tail #	CN	Lineage	Status	Last Unit	Date	Comments
52-1390	345	RB-36H RB-36H-II	Retired	5th BMW	27-May-57	To reclamation 24-Jun-57.
52-1391	347	RB-36H RB-36H-II	Retired	5th BMW	17-Jul-58	To reclamation 21-Jul-58.
52-1392	349	RB-36H RB-36H-II	Retired	28th BMW	15-May-57	To reclamation 29-Jul-57.
52-2210	351	YB-36J B-36J B-36J-III	Retired	95th BMW	4-Feb-59	B-36J prototype. First flew Jul-53. To reclamation 5-Feb-59.
52-2211	352	B-36J B-36J-III	Retired	95th BMW	14-Oct-58	To reclamation 15-Oct-58.
52-2212	353	B-36J B-36J-III	Retired	95th BMW	21-Jan-59	To reclamation 23-Jan-59.
52-2213	354	B-36J B-36J-III	Retired	95th BMW	29-Sep-58	To reclamation 30-Sep-58.
52-2214	355	B-36J B-36J-III	Retired	95th BMW	9-Dec-58	To reclamation 11-Dec-58.
52-2215	356	B-36J B-36J-III	Retired	95th BMW	27-Jan-59	To reclamation on 29-Jan-59.
52-2216	357	B-36J B-36J-III	Retired	95th BMW	16-Oct-58	To reclamation 17-Oct-58.
52-2217	358	B-36J B-36J-III	Retired	95th BMW	10-Feb-59	Flown to SAC Museum, Bellevue, NE on 22-Apr-59 for display.
52-2218	359	B-36J B-36J-III	Retired	95th BMW	10-Feb-59	To reclamation 12-Feb-59.
52-2219	360	B-36J B-36J-III	Retired	95th BMW	10-Dec-58	To reclamation 11-Dec-58.
52-2220	361	B-36J B-36J-III	Retired	95th BMW	3-Feb-59	Flown to Air Force Museum, Wright-Patterson AFB, OH on 30-Apr-59 for display. Last B-36 flight (non-operational).
52-2221	362	B-36J B-36J-III	Retired	95th BMW	5-Feb-59	To reclamation 10-Feb-59.
52-2222	363	B-36J B-36J-III	Retired	95th BMW	14-Oct-58	To reclamation 15-Oct-58.
52-2223	364	B-36J B-36J-III	Retired	95th BMW	10-Feb-59	To reclamation 12-Feb-59.
52-2224	365	B-36J B-36J-III	Retired	95th BMW	21-Jan-59	To reclamation 23-Jan-59.
52-2225	366	B-36J B-36J-III	Retired	95th BMW	15-Oct-58	To reclamation 17-Oct-58.
52-2226	367	B-36J B-36J-III	Retired	95th BMW	27-Jan-59	To reclamation 29-Jan-59.
52-2812	368	B-36J B-36J-III	Retired	95th BMW	16-Jan-59	To reclamation 20-Jan-59.
52-2813	369	B-36J B-36J-III	Retired	95th BMW	15-Oct-58	To reclamation 17-Oct-58.
52-2814	370	B-36J-III	Retired	95th BMW	5-Feb-59	Originally ordered as B-36J. Converted at factory and delivered as B-36J-III. To reclamation 10-Feb-59.

Tail #	CN	Lineage	Status	Last Unit	Date	Comments
52-2815	371	B-36J-III	Retired	95th BMW	23-Jan-59	Originally ordered as B-36J. Converted at factory and delivered as B-36J-III. To reclamation 26-Jan-59.
52-2816	372	B-36J-III	Retired	95th BMW	6-Feb-59	Originally ordered as B-36J. Converted at factory and delivered as B-36J-III. Second Broken Arrow involving a B-36 when a Mk-17 bomb was accidentally dropped during landing at Kirtland AFB, NM. To reclamation 10-Feb-59.
52-2817	373	B-36J-III	Retired	95th BMW	26-Sep-58	Originally ordered as B-36J. Converted at factory and delivered as B-36J-III. To reclamation 29-Sep-58.
52-2818	374	B-36J-III	Destroyed	6th BMW	25-May-55	Originally ordered as B-36J. Converted at factory and delivered as B-36J-III. Crashed near Sterling, TX when the aircraft disintegrated during thunderstorm or tornado (unknown). Wing and tail cone surfaces were blown off and the aircraft impact was almost straight down. Total 15 fatalities (all crew), 0 survivors.
52-2819	375	B-36J-III	Retired	95th BMW	16-Dec-58	Originally ordered as B-36J. Converted at factory and delivered as B-36J-III. To reclamation 17-Dec-58.
52-2820	376	B-36J-III	Retired	95th BMW	4-Feb-59	Originally ordered as B-36J. Converted at factory and delivered as B-36J-III. To reclamation 5-Feb-59.
52-2821	377	B-36J-III	Retired	95th BMW	10-Feb-59	Originally ordered as B-36J. Converted at factory and delivered as B-36J-III. To reclamation 13-Feb-59.
52-2822	378	B-36J-III	Retired	95th BMW	13-Jan-59	Originally ordered as B-36J. Converted at factory and delivered as B-36J-III. To reclamation 15-Jan-59.
52-2823	379	B-36J-III	Retired	95th BMW	13-Jan-59	Originally ordered as B-36J. Converted at factory and delivered as B-36J-III. To reclamation 15-Jan-59.
52-2824	380	B-36J-III	Retired	95th BMW	10-Feb-59	Originally ordered as B-36J. Converted at factory and delivered as B-36J-III. To reclamation 12-Feb-59.
52-2825	381	B-36J-III	Retired	95th BMW	3-Feb-59	Originally ordered as B-36J. Converted at factory and delivered as B-36J-III. To reclamation 5-Feb-59.
52-2826	382	B-36J-III	Retired	95th BMW	15-Jan-59	Originally ordered as B-36J. Converted at factory and delivered as B-36J-III. To reclamation 20-Jan-59.
52-2827	383	B-36J-III	Display	95th BMW	12-Feb-59	Originally ordered as B-36J. Converted at factory and delivered as B-36J-III. Displayed in Fort Worth, TX under Air Force Museum loan program until the late 1980s. Underwent several years of restoration beginning in 1992. Moved to Pima Air & Space Museum in Tucson, AZ in 2005.

Bibliography

Adams, C. (1999). *Inside the Cold War: A Cold Warrior's Reflection.* Maxwell AFB, AL: Air University Press.

Adler, L. (1994, January 20). Albuquerque's Near Doomsday. *Albuquerque Tribune.*

Baugher, J. (2000, September 19). *Convair B-36 Peacemaker.* Retrieved from USAF Bombers: http://joebaugher.com/usaf_bombers/b36.html

Boyne, W. J. (2007, May). Airpower Classics: B-36. *Air Force Magazine.*

Broken Arrow, A lost Nuclear (Fat Man Bomb). (n.d.). Retrieved from Mysteries of Canada: https://www.mysteriesofcanada.com

Correll, J. T. (2016, April). 1946: The Year After the War. *Air Force Magazine.*

Deaile, M. G. (2018). *Always At War; Organizational Culture in Strategic Air Command, 1946-62.* Annapolis, Maryland: Naval Institute Press.

Ford, D. (1996, April). B-36: Bomber at the Crossroads. *Air and Space Magazine.*

Hopkins, J. (1982). *The Development of Strategic Air Command 1946-1981 (A Chronological History).* Offutt AFB, NE: Office of the Historian, Headquarters Strategic Air Command.

Jacobsen, M. K. (1997). *Convair B-36: A Comprehensive History of America's "Big Stick".* Atglen, Pennsylvania: Schiffer Military History.

Jenkins, D. R. (2001). *Magnesium Overcast: The Story of the Convair B-36.* North Branch, MN: Speciality Press.

Knaack, M. S. (1988). *Post-World War II Bombers 1945-1973.* Washington DC: Office of Air Force History.

McGowan, S. (2016, October). The Craswell B-36 Disaster. *Air Force Magazine.*

O'Connell, J. (2013, August 13). *Fairchild Air Force Base's Legacy of Excellence - Operation Big Stick.* Retrieved from U.S. Air Force: fairchild.af.mil/News/Article

Philips, F. M. (1959, May). B-36 Deterrence...Yesterday's and Tomorrow's. *Air Force Magazine.*

Wack, F. J. (1990). *The Secret Expolorers: Sage of the 46th/72nd Reconnaissance Squadrons.* Self Published.

Wolk, H. S. (1988, May). Revolt of the Admirals. *Air Force Magazine.*

Notes and Citations

Design and Development

[1] By this time, the Soviet Union had made significant progress in long-range bomber design. In 1933 they launched the ANT-16, and in 1934 the ANT-20, which was the largest aircraft at that time. Both aircraft were larger than any American aircraft and were all-metal designs.

[2] Project A became the parent to the B-17, B-24, and B-29 which were employed heavily in World War II.

[3] Philips, F. M. (1959, May). *B-36 Deterrence... Yesterday's and Tomorrow's.* Air Force Magazine, p. 4

[4] Knaack, M.S. (1988). *Post World War II Bombers 1945-1973.* Washington DC: Office of Air Force History, p. 5

[5] (Knaack, 1988, pp. 5-6)

[6] Jenkins, D. R. (2001). *Magnesium Overcast: The Story of the Convair B-36.* North Branch, MN: Specialty Press, p. 4

[7] (Knaack, 1988, pp. 6-7)

[8] Designs were also submitted by Northrop based on the XB-35 Flying Wing, Martin based on the XB-33, and Douglas based on the XB-19. None of these resulted in a competitive design.

[9] (Philips, 1959, p. 5)

[10] Although commonly referred to as Convair from this point forward, the company was officially called Consolidated-Vultee Aircraft Corporation. The Convair name was not official until it became the Convair division of General Dynamics on 29 April 1954.

[11] (Philips, 1959, p. 5)

[12] Hap Arnold is the only Air Force officer to ever hold 5-star rank. He was promoted to General of the Army on 21 December 1944 and his rank was posthumously redesignated General of the Air Force on 7 May 1949.

[13] Ford, D. (1996, April). *B-36: Bomber at the Crossroads.* Air and Space Magazine, p. 3

[14] Correll, J. T. (2016, April). *1946: The Year After the War.* Air Force Magazine, pp. 65-66

[15] (Correll, 2016, p. 69)

[16] (Philips, 1959, p. 1)

[17] (Philips, 1959, p. 2)

[18] (Philips, 1959, p. 2)

[19] (Philips, 1959, p. 2)

Roles and Missions

[1] Hopkins, J. (1982). *The Development of Strategic Air Command 1946-1981 (A Chronological History).* Offutt AFB, NE: Office of the Historian, Headquarters Strategic Air Command, p. 2

[2] (Ford, 1996, p. 4)

[3] (Ford, 1996, p. 4)

[4] Wolk, H. S. (1988, May). *Revolt of the Admirals.* Air Force Magazine, p. 2

[5] (Wolk, 1988, p. 3)

[6] (Wolk, 1988, p. 3)

[7] (Wolk, 1988, p. 4)

[8] (Ford, 1996, p. 1)

[9] (Wolk, 1988, p. 4)

[10] (Wolk, 1988, p. 4)

[11] (Ford, 1996, p. 2)

[12] (Ford, 1996, p. 2)

[13] (Wolk, 1988, p. 5)

[14] (Wolk, 1988, p. 5)

[15] (Wolk, 1988, p. 5)

[16] Adams, C. (1999). *Inside the Cold War: A Cold Warrior's Reflection.* Maxwell AFB, AL: Air University Press, p. 37

[17] Wack, F. J. (1990). *The Secret Expolorers: Sage of the 46th/72nd Reconnaissance Squadrons.* Self Published.

[18] Jacobsen, M. K. (1997). *Convair B-36: A Comprehensive History of America's "Big Stick".* Atglen, Pennsylvania: Schiffer Military History.

[19] Wikipedia RB-36

Models and Variants

[1] (Ford, 1996, p. 4)

[2] Originally codenamed SILVERPLATE but changed to SADDLETREE in 1947 after the SILVERPLATE codename was compromised.

[3] (Jenkins, 2001, p. 147)

[4] (Jenkins, 2001, p. 147)

[5] (Jenkins, 2001, pp. 148-149)

[6] (Jenkins, 2001, p. 145)

[7] (Knaack, 1988, p. 7); (Jenkins, 2001, pp. 7-8)

[8] (Knaack, 1988, p. 12)

[9] (Jenkins, 2001, pp. 6-8)

[10] (Jenkins, 2001, p. 16)

[11] Boyne, W. J. (2007, May). *Airpower Classics: B-36.* Air Force Magazine, p. 168

[12] Not to be confused with Major General Thomas Power who was Deputy Commander of SAC under LeMay and would later replace LeMay as CINCSAC.
[13] (Jenkins, 2001, p. 37)
[14] This aircraft and the next 12 B-36A aircraft were originally designated YB-36A and then re-designated B-36A before delivery. The aircraft historical record indicates this aircraft was delivered as a B-36A and retained this designation until it was destroyed. However, most historical accounts indicate it was designated YB-36A.
[15] (Knaack, 1988, p. 18)
[16] Now called Wright-Patterson AFB.
[17] The aircraft was later returned to B-36A configuration briefly before being converted to RB-36E.
[18] (Knaack, 1988, p. 18)
[19] According to most accounts, B-36A (44-92004) was the first B-36A delivered to the 7th BG at Carswell AFB. However, this aircraft was flown directly to Wright Field for testing to destruction. Also, accounts show that the first aircraft was delivered to Carswell on 26 June 1948. However, Air Force historical records indicate this occurred on 28 June 1948.
[20] (Ford, 1996, p. 5)
[21] The Tarzon bomb was an early guided bomb developed by combining the VB-3 Razon and the British Tallboy 12,000-pound earthquake bomb resulting in the Tallboy Range and Azimuth Only bomb. It was used on the B-29 during the Korean War but retired in 1951. It was never used on the B-36.
[22] (Ford, 1996, p. 5)
[23] (Jenkins, 2001, pp. 63-65)
[24] This aircraft was delivered as an RB-36D on 30 October 1950.
[25] (Jenkins, 2001, p. 64)
[26] (Ford, 1996, pp. 5-6)
[27] (Knaack, 1988, p. 31)
[28] It is unclear if 49-2647 is the correct tail number that first flew on 11 July 1949. However, historical accounts indicate the first "true production B-36D" flew that day so it is reasonable to conclude it is the correct tail number.

[29] Historical accounts show differing dates for the official first flight including November 1951 and 5 April 1952. However, Air Force historical records indicate the aircraft was available on 5 November 1951 and delivered on 31 December 1951, so it is reasonable to assume the aircraft actually flew for the first time in November 1951.
[30] Baugher, J. (2000, September 19). *Convair B-36 Peacemaker.* Retrieved from USAF Bombers: http://joebaugher.com/usaf_bombers/b36.html
[31] United States Air Force RB-36F Factsheet
[32] United States Air Force RB-36H Factsheet
[33] The first flight date is not documented in historical records. However, the first aircraft delivered (50-1104) became available on 29 November 1951 and was accepted on 13 January 1952 so it is reasonable to assume the first flight occurred in January 1952.
[34] (Philips, 1959, p. 7)
[35] Convair assumed the B-36C prototype would be successful and the Air Force would subsequently order more aircraft. They submitted a proposal for additional production aircraft on 5 May 1947.
[36] (Jenkins, 2001, pp. 71-72)
[37] Wikipedia RB-36
[38] (Jenkins, 2001, p. 92)
[39] United States Air Force GRB-36F Factsheet
[40] (Philips, 1959, p. 7)
[41] (Jenkins, 2001, p. 79)
[42] (Ford, 1996, pp. 8-9)

Organization and Basing

[1] (Correll, 2016, p. 67)
[2] (Correll, 2016, p. 65)
[3] SAC also established the 4th AD at Barksdale AFB, 6th AD at MacDill AFB, 12th AD at March AFB, 14th AD at Travis AFB, and 47th AD at Walker AFB on 10 February 1951, 5th AD in French Morocco on 14 June 1951, and 7th AF in England on 20 March 1951. It inherited 3rd AD on Guam when FEAF was discontinued on 18 June 1954.
[4] (Jenkins, 2001, p. 48)
[5] Later the Arch Hangar became Loring's phase dock and housed one B-52 and two KC-135 aircraft during inspections.
[6] (Knaack, 1988, pp. 50-51)
[7] (Philips, 1959, p. 1)

First and Records

[1] (Jenkins, 2001, p. 43)
[2] (Jenkins, 2001, pp. 43-44)
[3] (Jenkins, 2001, p. 44)
[4] (Knaack, 1988, p. 23); (Jenkins, 2001, p. 44)
[5] (Jenkins, 2001, p. 44)
[6] (Jenkins, 2001, p. 44)
[7] (Hopkins, 1982, p. 17); (Jenkins, 2001, p. 52)
[8] (Jenkins, 2001, p. 92)

Operations

[1] (Jenkins, 2001, p. 52)
[2] LeMay's title changed twice during his tenure in SAC. It was changed from Commanding General to Commander in 1953 and then to Commander-In-Chief Strategic Air Command (CINCSAC) in 1955. He was promoted to General on 29 October 1951 becoming the youngest 4-star since General Grant.
[3] (Ford, 1996, p. 5)
[4] (Hopkins, 1982, pp. 16-17)
[5] (Hopkins, 1982, p. 18)
[6] (Ford, 1996, p. 6)
[7] (Jenkins, 2001, p. 48)
[8] (Jenkins, 2001, p. 44)
[9] (Jenkins, 2001, p. 48)
[10] (Jenkins, 2001, p. 51); (Knaack, 1988, p. 24)
[11] (Knaack, 1988, p. 24); (Jenkins, 2001, p. 51); B-36B Factsheet
[12] A B-50 made a similar flight of 41.7 hours covering 9,870 miles with three in-flight refueling by KB-29 tankers.
[13] (Knaack, 1988, p. 24)
[14] (Jenkins, 2001, p. 48)
[15] (Knaack, 1988, p. 24); (Jenkins, 2001, p. 51)
[16] (Knaack, 1988, p. 48)
[17] (Boyne, 2007, p. 168)
[18] (Jenkins, 2001, pp. 51-53)
[19] (Jenkins, 2001, p. 53)
[20] (Knaack, 1988, p. 32); (Hopkins, 1982, p. 33)
[21] (Hopkins, 1982, p. 44)
[22] (O'Connell, 2013)
[23] (Knaack, 1988, p. 35); (Hopkins, 1982, p. 49)
[24] (Jenkins, 2001, p. 71)
[25] (Jenkins, 2001, p. 96)
[26] (Hopkins, 1982, p. 14)
[27] (Hopkins, 1982, p. 18); (Jenkins, 2001, p. 52)
[28] (Hopkins, 1982, p. 33)
[29] (Hopkins, 1982, p. 34)
[30] (Hopkins, 1982, p. 39)
[31] (Hopkins, 1982, p. 44)
[32] (Hopkins, 1982, pp. 49-50)
[33] (Hopkins, 1982, pp. 54-55)
[34] (Hopkins, 1982, p. 58)
[35] (Hopkins, 1982, p. 68)
[36] (Hopkins, 1982, p. 74)
[37] (Jenkins, 2001, p. 243)
[38] (Jenkins, 2001, p. 244)
[39] (Jenkins, 2001, p. 244)
[40] (Jenkins, 2001, p. 245)
[41] (Jenkins, 2001, p. 245)
[42] (Jenkins, 2001, p. 245)
[43] (Jenkins, 2001, p. 247)
[44] (Jenkins, 2001, p. 247)
[45] (Jenkins, 2001, p. 247)
[46] (Jenkins, 2001, p. 248)
[47] (Jenkins, 2001, p. 248)
[48] (Jenkins, 2001, p. 248)
[49] (Jenkins, 2001, pp. 248-249)
[50] (Jenkins, 2001, p. 249)
[51] (Jenkins, 2001, p. 249)
[52] McGowan, S. (2016, October). *The Craswell B-36 Disaster.* Air Force Magazine, p. 64
[53] The damage that occurred at Carswell in 1952 is now widely believed to have been from a microburst rather than a tornado. A microburst was also blamed for the crash of a Delta Airlines L-1011 at Dallas-Fort Worth airport only 30 miles from Carswell in 1985.
[54] (McGowan, 2016, p. 65)
[55] Broken Arrow is the term used to describe an accident involving a nuclear weapon.
[56] *Broken Arrow, A lost Nuclear (Fat Man Bomb),* (n.d.) Retrieved from Mysteries of Canada: https://www.mysteriesofcanada.com
[57] Adler, L. (1994, January 20). *Albuquerque's Near Doomsday.* Albuquerque Tribune, p. 2

Displays

[1] By this time, the Air Force had about 500 B-52 aircraft fielded.
[2] Now called National Museum of the United States Air Force.
[3] The museum's name was changed to Strategic Air & Space Museum in 2001 and then to Strategic Air Command & Aerospace Museum in 2015.

About the Author

H.J. Campbell is a long-time veteran of the U.S. Air Force and the defense aerospace industry. He completed 20 years in the Air Force including 15 years in SAC. His SAC assignments included the 42nd BMW at Loring AFB, ME, 320th BMW at Mather AFB, CA, 43rd Strategic Wing at Andersen AFB, Guam, and HQ SAC Maintenance Standardization and Evaluation Team (MSET). After SAC was deactivated in Jun 1992, he was transferred to HQ Air Mobility Command (AMC) where he supported much needed KC-135 avionics upgrades. After his retirement from the Air Force, he served in the aerospace industry as a program manager and director for modification, maintenance, and logistics programs on KC-135, KC-10, C-130, C-27, and other military aircraft. With this book, the first in a planned series about SAC aircraft, he is embarking on his "third career" as a military historian and author.

Also available from *Electrikbooks* at Amazon.com, BarnesandNoble.com, and BooksaMillion.com

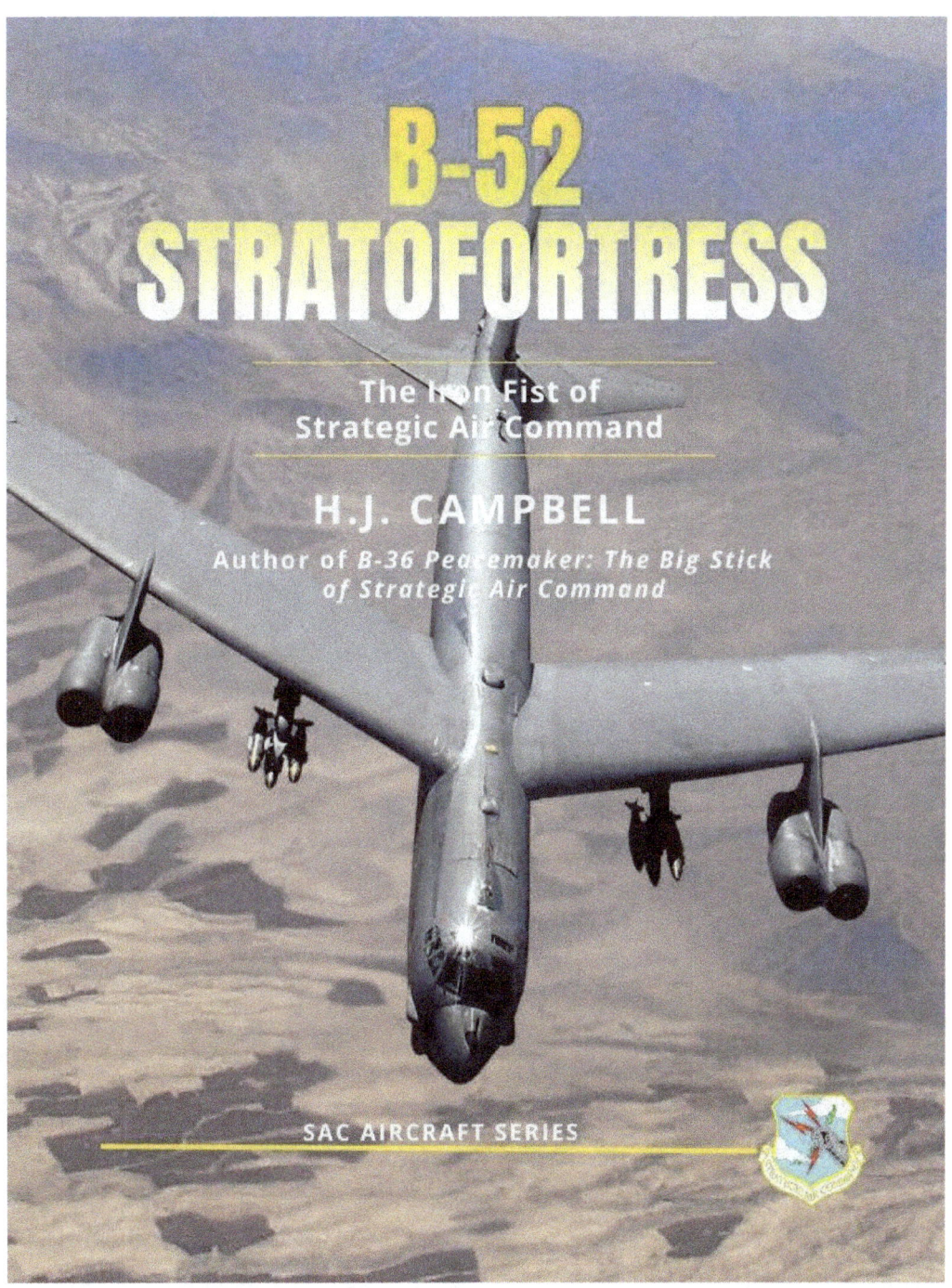

www.ingramcontent.com/pod-product-compliance
Lightning Source LLC
Chambersburg PA
CBHW082209070526

44585CB00020B/2339